Promiscuity

by the same author

ATLANTIC ALCIDAE, SPERM COMPETITION IN BIRDS, THE MAGPIES, GREAT AUK ISLANDS, SPERM COMPETITION AND SEXUAL SELECTION, AVIAN ECOLOGY, THE SURVIVAL FACTOR

Promiscuity

An Evolutionary History of Sperm Competition

TIM BIRKHEAD

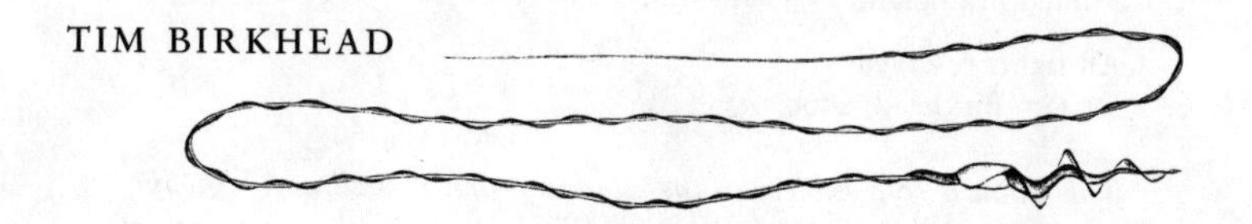

Harvard University Press
Cambridge, Massachusetts
2000

First published in 2000
by Faber and Faber Limited
3 Queen Square London WC1N 3AU

Photoset by Wilmaset Ltd, Birkenhead, Wirral
Printed in England by Clays Ltd, St Ives plc

Library of Congress Cataloging-in-Publication Data

Birkhead, T. R.
Promiscuity : an evolutionary history of sperm competition / Tim Birkhead.
p. cm.
Includes bibliographical references (p.).
ISBN 0-674-00445-0
1. Sexual selection in animals. 2. Sperm competition. 3. Reproduction
4. Promiscuity
I. Title.
QL761 .B57 2000
591.56'2—21

00-040923

Contents

List of Plates

1 Computer-enhanced images by O. D'Cruz.
Photographs: 2 (a) R. Snook; (b) H. D. M. Moore; (c) W. G. Breed; (d) C. W. LaMunyon. 3 (a) M. D. Norman; (b) N. Wedell; (c) T. R. Birkhead. 4 S. Pitnick. 5 Claude Carré, Danielle Carré, Evelyn Houliston, Christian Rouvière and Christian Sardet. 6 (a) R. Lefebvre, H.-C. Ho and S. Suarez; (b) T. R. Birkhead; (c) T. Karr and R. Snook; (d) P. Sutovsky. 7 (a) and (b) A. F. Dixson; (c) T. R. Birkhead; (d) R. Wilkinson. 8 (a) M. Jennions; (b) S. Neilsen; (c) P. Ward; (d) A. Syred; (e) H. Hoffer; (f) N. Michiels.

Preface

This is a book about reproduction. It is about the causes and consequences of female promiscuity, and in particular the ways in which the two components of Darwin's concept of sexual selection – competition between males and choice by females – operate after insemination has taken place. Post-copulatory sexual selection, as it is called, consists of competition between the sperm of different males to fertilize females' eggs (sperm competition) and choice of different males' sperm by females (sperm choice). These are, by definition, processes that can occur only if females are inseminated by more than one male during a single reproductive cycle, and their effects are far-reaching, shaping not only the physiological events that occur after insemination, but also many behaviours that occur before and during copulation. Generations of reproductive biologists assumed females to be sexually monogamous but it is now clear that this is wrong. The very recent recognition that females of most species are promiscuous and routinely copulate with several different males, together with the realization that in an evolutionary sense all organisms are basically selfish, has revolutionized our view of reproduction.

The traditional view of reproduction was summed up by the writer Gerald Brenan[1] in the following way: 'Since copulation is the most important act in the lives of living creatures because it perpetuates the species, it seems odd that Nature should not have arranged for it to happen more simply.'

Obviously he is correct in recognizing the fundamental necessity of copulation in reproduction, but Brenan's 'puzzlement' over the lack of ease with which copulation occurs rests on the erroneous assumption that reproduction is a co-operative venture between males and females and that as such it serves to perpetuate the species. Although Brenan's comment is based on his own experience, it nevertheless identifies an important general point: that there often exists a conflict of interest between males and females over copulation. It is no longer meaningful to consider reproduction, whether it involves copulation, or simply the release of gametes into the sea, as a collaboration between the sexes. More accurately, it is a potent mix of competition between males and choice by females which together generate sexual conflict. Recent research has provided dramatic demonstrations that reproduction occurs neither for the good of the species nor as a mutually beneficial interaction between males and females. For example, as a consequence of female promiscuity, males often inflict damage on females in their competition to fertilize their eggs. Conversely, females sometimes inflict damage on males in the process of using the sperm of one male in preference to those of another. The damage the males of some species can inflict on females is insidious. As well as carrying sperm, their semen also contains substances that, on entering the female's bloodstream (via her reproductive tract), affect her brain and cause her to behave in a way that increases the male's reproductive success. As if this wasn't enough of a sexual conflict, these chemicals also reduce the female's lifespan. In other instances, females have the upper hand, and by encouraging males to copulate with them they acquire a ready source of nutrients, which channelled into eggs and babies enhances their own reproductive success. Sexual reproduction is anything but co-operative.

The scientific study of reproduction, particularly the physiology of reproduction, is relatively recent compared with that of other important body systems. One reason for this delay is the association between reproduction and sexuality, and the fact that for a long time reproductive biology was not considered a respectable topic for scientific enquiry.[2] However, the combined efforts of many individuals working in three rather disparate fields – biology, agriculture and medicine – mainly during the past hundred years, have helped to make the study of reproduction a more socially acceptable enterprise, as I shall show. However, the study of sexuality – that is, human sexuality – continues to be controversial even in the liberal last decades of the twentieth century, as evinced by the lack or withdrawal of governmental funding for studies of human sexual behaviour in efforts better to understand AIDS.[3] Notwithstanding official disapproval, as individuals we know that sexuality is at the very core of human being – inter-sexual relationships form the fabric of society. And herein lies a difficulty: the very fact that sex is so important to us means that we are vulnerable to being exploited by it. This, in turn, means that it is important that we understand it – and particularly from an evolutionary perspective.

Both sexes can be promiscuous, but while males are renowned for their promiscuity, until very recently females were not. The discovery that females of almost all animal species routinely copulate with several different males has, as I shall show, had a profound effect on the study of reproduction. Unfortunately, to describe a female, and in particular a human female, as promiscuous can imply a form judgement – in part because to describe a man as promiscuous does not, at least not in the same way. While biologists may be happy to refer with an air of scientific objectivity to either sex behaving promiscuously, when it comes to people it is much more difficult to remain objective. Since this book is primarily about female promiscuity I have sought an unemotive expression to use in the text to refer to females copulating with several males. The term 'polyandry' – literally, 'many males' – seems to me to be the best of those available since it is relatively neutral and has been used by others in this way. Polyandry is sometimes used to describe a mating system in which a female forms social and sexual relationships with several males. To avoid confusion I shall make this distinction explicit by referring to the latter as a polyandrous mating system. The expression 'multiple mating' has sometimes also been used by scientists to describe females copulating with several different males, but this is potentially confusing since females can copulate multiply with either the same or different males – here we are interested mainly in the latter situation. Moreover, the terms 'mating' and 'mate' are in themselves potentially ambiguous.

Over the past fifteen years I have met and got to know most of the practitioners in the field of sperm competition. Many of them have become good friends and with some of them I have also collaborated. My main acknowledgement is to all my colleagues in this field – without them there would have been little to write about. I have relied extensively on their published work and in order to clarify my own ideas I pestered some of them relentlessly as I wrote this book. I am especially grateful to those who provided photographs. Not all of the following are reproductive biologists, but they all took time to respond to my queries and I am extremely grateful to them: Björn Afzelius, Bruno Baur, Mike Bedford, John Bishop, Janet Browne, Helena Cronin, Jim Cummins, Alan Dixson, Bill Eberhard, Paul Eady, Matt Gage, Pascal Gagneux, Tim Glover, Karen Green, Sandy Harcourt, Ben Hatchwell, Kristen Hawkes, Dave Hosken, Lotta Kvarnemo, Peter Lake, Helmut Lemke, Kate Lessells, Ruth Mace, Janet Kear, Bart Kempenaers, Robin McCleery, Claire Lovell-Mansbridge, John Manning, Garry Marvin, Robert Mason, Nico Michiels, Anders Møller, Harry Moore, James Moore, Paulo Gama Mota, Mats Olsson, Jim Overstreet, Allan Pacey, Geoff Parker, Klaus Peschke, Scott Pitnick, Tomasso Pizzari, Percy Rhode, Martha Robbins, Gunilla Rosenqvist, Robert M. Sapolsky, Andrew M. Shedlock, Roger V. Short, Pascale Sicotte, Leigh Simmons, Eric A. Smith, Manolo Soler, Alistair Stutt, Jonathan Waage, Paul Ward, David Westneat and Graham Wishart. Apologies for anyone I have overlooked. I also owe thanks to my colleagues in the Department of Animal and Plant Sciences at the University of Sheffield: David Hollingworth for his technical skill with the colour images, and for stimulating discussion and banter, Terry Burke, Ben Hatchwell, Len Hill, Francis Ratnieks, Mike Siva-Jothy and Penny Watt.

Over the past few years much of my own research has been conducted in collaboration with Bobbie Fletcher and Jayne Pellatt; their help and companionship are greatly appreciated. I am especially grateful to Jayne Pellatt for drawing the figures.

I am particularly indebted to Andrew Balmford, Ben Hatchwell, Craig LaMunyon, Geoff Parker, Roger Short, Robert L. Smith and Richard Wagner, as well as Luke Vinten and Julian Loose at Faber and Faber for reading and commenting on the manuscript. Their suggestions were sometimes funny, often critical, but always instructive. Any errors are my own. Finally, I thank my family for their forbearance, and my children in particular, who were incredulous that such a small book could occupy me for so many evenings and weekends.

1 Competition, Choice and Sexual Conflict

The cuckold is the last that knows it.
OLD ENGLISH PROVERB

The shooting of migratory birds has long been an important part of Mediterranean culture. As with hunters everywhere, a male's success as a hunter is intimately linked with his status and, ultimately, with his reproductive success. Status for the Mediterranean male is all-important, and tradition dictates that a man who fails during a hunting expedition can expect his wife to be unfaithful. In parts of Italy it is widely believed that a man must shoot a honey buzzard each year if his wife is to remain faithful.[1] So strong is this belief, and so powerful a motivating force is the idea of female fidelity, that even after they have emigrated to the United States many Italian men return home each year to shoot a honey buzzard. It is not a little ironic that in order to fulfil this ritual a man usually leaves his wife behind. Moreover, in some instances it is the wife who actually encourages him to do so! As we shall see, for a male to leave his partner unattended is a risky business.

Concern about the paternity of his offspring is deeply embedded in the male psyche or, rather, in his genes. There would be no concern about paternity if females were always faithful, but they are not. Given that we are all aware of infidelity and its consequences, it is rather odd that for much of the time we choose to ignore it. The popular image of how each of us got started is a single large ovum surrounded by a wriggling mass of tiny sperm, each attempting to achieve fertilization. An implicit assumption is that all those sperm come from the same man. For almost all organisms, including ourselves, this is wrong. In most instances the sperm from different males compete to fertilize a female's eggs. This is sperm competition. How could this occur?

What does it mean for the female within whose oviduct this scenario is enacted? And what does it mean for the males that inseminated the female? Exactly how and why the sperm of two males end up in the vicinity of a female's eggs just at the point of fertilization form the basis of this book.

The risk of cuckoldry, the drive to cuckold and the need to reproduce has opened not just a single Pandora's box, but a whole set of them. The recent realization that sperm competition is widespread across the animal kingdom, from flatworms to flounders and from honey bees to humans, has changed the way biologists view the world. Not only that, the existence of sperm competition as an evolutionary force has helped us to understand bits of biology, including our own, that were previously inexplicable. Its explanatory power is so enormous we might wonder why it has taken so long to realize the importance of sperm competition.

For several centuries before the birth of Christ, Aristotle and his predecessors were aware of the existence of sperm competition. They knew, for example, that if a bitch copulated with two dogs during a single heat she could produce pups fathered by each male. Indeed, exactly this observation might have provided the basis for the story of Heracles and Iphicles: the night before Amphitryon's marriage to Alcmene, Zeus disguised himself as Amphitryon and slept with her. The next night Amphitryon consummated his marriage and in due course Alcmene produced twins: Iphicles (fathered by Amphitryon) and Heracles (fathered by Zeus).[2]

Although the Greeks knew about multiple paternity and that the transfer of semen from male to female was part of reproduction, they thought that insemination was an incidental process in which the male simply provided the soul for the new being.[3] The female's obvious and nourishing role in pregnancy was paramount. Part of the confusion stemmed from an asymmetry of knowledge about male and female reproductive anatomy. The reproductive equipment of the male was well known since war provided an abundant supply of male bodies for physicians to dissect, but the dissection of women was forbidden. Aristotle was probably the first accurately to illustrate the male urino-genital system, but his knowledge of female anatomy was abysmal. It was not until four hundred years after Aristotle's death that the reproductive system of the human female was illustrated correctly – and even now we know precious little about the function of parts of it, such as the clitoris. Aristotle was also confused by menstruation. He thought that females produced semen, from their testes (= ovaries) and that the human egg was produced in the womb from the interaction between menstrual blood and the female's semen. Four hundred years or so later Galen (AD 130–201), chief physician to the gladiators, suggested that the female testes produced a semen which mixed with male semen in the uterus to form a *coagulum* which gave rise to an embryo. This was essentially correct, although several more centuries would pass before the full facts were established.

Aristotle wrote extensively about reproductive behaviour and anatomy, but his reputation has fluctuated over the years. Peter and Jean Medawar (1984) described him as a careful collector and observer of an enormous range of facts, but not a scientist.[4] His 'biological works ... are a strange and generally speaking rather tiresome farrago of hearsay, imperfect observation, wishful thinking, and credulity amounting to downright gullibility ... Sometimes Aristotle is right; his writings were so voluminous he could hardly fail to be correct sometimes (irreverent thoughts of monkeys and typewriters steal into the mind).'

Contrary to the impression given by the Medawars, Aristotle was correct about quite a lot, including sperm competition. He made numerous observations of copulation and egg formation in chickens in an attempt to understand the timing and process of fertilization. In particular he noticed that when a hen copulated with two cockerels of different types the resulting offspring usually resembled the second male to mate. In this single observation Aristotle provides an explicit statement of the occurrence of sperm competition, but also, by using what we would now call genetic markers, was able to ascertain the paternity of the chicks and record the occurrence of a phenomenon we now refer to as last male sperm precedence.[5]

That the Medawars' sweeping generalization about Aristotle was unjust has been nicely demonstrated by Robin Dunbar, now at the University of Liverpool. He showed that Aristotle was generally right about those things he could observe directly, and often wrong about those he could not.[6] As we shall see, Aristotle's reference to sperm precedence in the domestic fowl turned out to be correct.

Darwinian Origins

Despite this auspicious start, little of much significance happened in terms of sperm competition during the two thousand years following Aristotle's death. That is not to say that nothing was discovered, for it was during this period that most of the fundamental aspects of reproduction were established, including the existence of sperm and ova and their fusion to form embryos. Only in the nineteenth century, with the emergence of Charles Darwin as the most influential biologist ever, did the idea of sperm competition resurface. Once Darwin had identified natural selection as the mechanism by which evolutionary change occurs, he turned his mind to the differences between males and females. The puzzle was this: if natural selection favoured those individuals which survived to leave lots of offspring, how could structures or behaviours that were so obviously detrimental to their bearers evolve by natural selection? What Darwin was thinking of here was the gaudy plumage or songs of male birds which made them conspicuous to predators, and the huge antlers of stags which required considerable effort to grow and to carry. The answer to this question had already been anticipated, at least in part, by Darwin's grandfather. In his book *Zoonomia*, Erasmus Darwin discussed how horns, spurs and other attributes had evolved to enable males to fight others for the possession of females. As a young student in Edinburgh, Charles had read his grandfather's tome and although Erasmus is fairly explicit about the evolution of male fighting traits, for some reason Charles never gave him much credit for these ideas. While it was fairly obvious that many of the morphological and behavioural features that distinguished males from females resulted from competition between males for females, Charles also considered the numerous male traits, like bright feathers and wattles, that could have nothing to do with male fighting ability. To explain these features he suggested that females had a sense of beauty and preferred to copulate with attractive males, although he wasn't exactly clear why. Darwin therefore saw the evolution of sexual differences resulting from two processes: competition between males and choice of males by females. Together these two processes comprised what he called sexual selection.[7] Although virtually neglected in the decades immediately following its origin, by the 1930s Darwin's concept of sexual selection was acknowledged as being on a par with that of natural selection by evolutionary biologists and, as Michael Ghiselin[8] pointed out in 1969, was 'his most brilliant argument in favour of natural selection'.

Darwin himself was under no illusions about the significance of sexual selection. He knew that it provided an all-encompassing explanation for many of the differences between the sexes. However, I think Darwin would have been delighted by the way the study of sexual selection blossomed into one of the main fields of evolutionary biology by the end of the twentieth century. What Darwin might not have anticipated was the extent to which his clever idea would subsequently be developed. When he first proposed sexual selection as a process he assumed that it operated only up to the point where an individual had acquired a mate. The fights between males were for possession of females. The choosiness of females was concerned with the decision of who to pair up with, or who to copulate with. As Darwin saw it, once either sex had acquired a partner that was the end of it. This view of sexual selection was to persist for another hundred years until two young biologists, on opposite sides of the Atlantic, burst on to the scene: Robert Trivers and Geoff Parker.

Parental Investment

At Harvard University in the late 1960s Robert (Bob) Trivers was rethinking sexual selection. Just like Darwin, Trivers was interested in what determines the intensity of sexual selection or, put another way, why some individuals are so much more successful in reproduction than others. The greater the difference in reproductive success between the most and the least successful individual the more intense is sexual selection. Trivers's revolutionary idea was that the amount of effort that each sex put into producing and rearing offspring, or parental investment as he called it, determines sexual selection's intensity. Trivers's novel contribution was the suggestion that whichever sex puts least into rearing offspring shows the greatest variation in reproductive success.[9] To spell it out: sexual selection operates most intensively on the sex that invests least in rearing offspring.

In most species males invest less in reproduction than females. Males sometimes contribute nothing more than a drop of semen. On a cell-by-cell basis a male's sperm are much smaller than a female's eggs or ova. In addition, females may carry the developing foetus for months and then may continue to provide care and protection for weeks or even years after offspring are born. Only rarely do males provide as much, or even more, care than females. Trivers pointed out that species varied in the relative parental investment made by males and females and that this difference often coincided with the difference in the appearance of the sexes. His explanation for this pattern was that the sex that invests most becomes limiting for the other sex which then has to compete even more furiously for those partners still available.

The easiest way to visualize this is to imagine a population of peafowl with equal numbers of males and females. The plumage and behaviour of the male and female are as different as their investment patterns. Peacocks provide a few million sperm and nothing more, but peahens produce and incubate a clutch of energy-rich eggs and care for their chicks for several weeks after hatching. Once a female has been inseminated she is out of the game – she leaves the area where males display and seeks solitude to rear her offspring. In contrast, within minutes of inseminating a female, a peacock is ready to go again, and because he has no family commitments he remains ready to copulate for the entire breeding season. The upshot is that sexually active males almost always outnumber sexually receptive females. In turn this then leads to intense competition among males for the ever decreasing pool of receptive females. At the same time, in this buyers' market, females can afford to be selective about whom they copulate with. The outcome is that while all females are inseminated and produce offspring, the reproductive success of males is highly variable: the most aggressive or most attractive father many offspring, while others may attain no reproductive success whatsoever. The two processes of male competition and female choice favour those males which can fight and look good to females – hence the peafowl's sexual dimorphism.

Contrast peafowl with magpies. Male magpies invest huge amounts of energy in reproduction: they copulate (of course), but in addition they feed the incubating female and when the young hatch the male also provides most of their food. Moreover, a male magpie's parental duties limit his ability to inseminate other females. The end result is similar reproductive success among males and, as a consequence, very little sexual dimorphism: male and female magpies could hardly look more similar.

By considering the exceptions, Bob Trivers provided convincing evidence for the idea that it was the level of investment by each sex, and not gender itself, that determined the intensity of sexual selection. In a group of shore birds known as phalaropes the roles of the sexes are reversed and males undertake the main parental duties of incubation and caring for the young. In this situation females compete for males and males choose between females. Sexual selection therefore operates more strongly on females, who, as a consequence, are larger, more aggressive and more colourful than males.

Trivers's revolutionary idea had its origins in a study conducted some twenty years previously by Angus Bateman[10] on the fruitfly *Drosophila melanogaster*. It seems unlikely that the behaviour of the humble fruitfly should provide a model for how animals, including ourselves, behave. But this is exactly what Trivers succeeded in demonstrating. Bateman used genetic markers, such as differences in eye colour, whose mode of inheritance was known, to measure the reproductive success of both male and female flies. In this way he ingeniously obtained the type of results we would get today by using molecular techniques to assign parentage. In each experiment Bateman placed three or four males with three or four females in a vial for three or four days, after which he allowed the resulting eggs to hatch and develop so he could figure out who had copulated with whom. Overall, Bateman determined the mothers and fathers of no fewer than 9,500 flies. His results revealed a dramatic difference between males and females. For a male, the more females he copulated with, the greater his reproductive success. For females, on the other hand, one copulation was enough. As long as a female was inseminated once she had enough sperm to fertilize her entire complement of eggs. Additional copulations did nothing to increase her reproductive success.

Bateman concluded that male fruitflies do better the more they copulate, but females need to copulate only once and gain rather little from copulating any more (figure 1). Bateman recognized that his study had broad implications and suggested that his results should apply very widely, to 'all but a few primitive organisms, and those in which monogamy combined

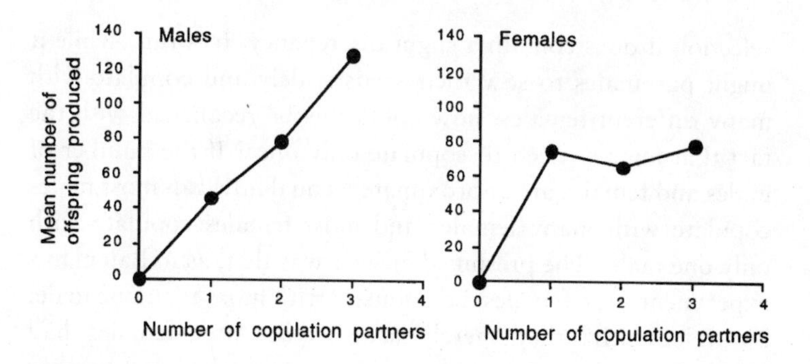

Figure 1

Results from Angus Bateman's experiments on *Drosophila* showing that male reproductive success increases with the number of copulation partners, whereas female reproductive success does not increase by copulating with more than one male (from Arnold and Duvall (1994))

with a sex ratio of unity eliminated all intra-sexual selection'. Trivers championed Bateman's work, expanded it and forged it into one of the foundation stones of a new field of study, now called behavioural ecology. Trivers's ideas subsequently formed the basis for an entire generation of research on sexual selection and sperm competition. He also introduced the concept of a 'mixed reproductive strategy', in which males of socially monogamous species, like birds, form a pair bond with one female but do not pass up the opportunity to copulate with others because such behaviour would further enhance male reproductive success. One reason, Trivers suggested, why males were able to behave in this way was that, unlike females, their reproductive success was not limited by their ability to produce sperm. Individual sperm were much smaller and hence cheaper to produce than eggs, and males seemed to have an almost limitless supply. These ideas were consistent with the widespread view that sexual selection operated more intensively on males than on females.

Despite the inherent appeal of Trivers's model of sexual

selection it does contain a slight discrepancy. It is this: while it might pay males to sow their seeds widely and copulate with many different females, how could this be reconciled with the fact that females need to copulate only once? If the number of males and females are approximately equal how can most males copulate with many females and most females copulate with only one male? The presumed answer was that, as in Bateman's experiments, the females did copulate with more than one male, but reluctantly: they merely acquiesced. After females had copulated with one male, on being approached by another they lay back and thought of England, but their hearts weren't in it. For a while at least this view of female behaviour satisfied most biologists but, as we shall see, it was flawed.

The Father of Sperm Competition

At the same time that Trivers was developing his ideas about sexual selection, Geoff Parker, now at Liverpool University, was thinking about one particular aspect of sexual selection: sperm competition. As an undergraduate at Bristol, Parker had been encouraged to study the behaviour of the yellow dungfly by the eminent entomologist Howard Hinton. As was true of many able students in those heady days in higher education, Parker was persuaded to continue this work for a Ph.D., still under Hinton's supervision. Only following a year of directionless study did Parker realize that if he was going to achieve anything he needed to address a specific question. After spending several months with his nose a few centimetres from fresh cowpats observing the intense competition between male dungflies for females, Parker recognized that the problem he should address was sexual selection – and specifically male–male competition. The issue was this: a female dungfly arrives at the cowpat ready to copulate and is then almost immediately ready to lay her eggs in the surface of the dung. She doesn't have to wait long, for as soon as she alights she is grabbed by a male, who instantly copulates with her. Before he has finished, however, a larger male grasps the pair, rips the first male from the female, discards him and copulates with the female. This scenario was enacted hour after hour, day after day, on every fresh dungpat. You can go out on a warm summer day and witness it for yourself: female dungflies routinely copulate with several males. Geoff Parker realized that when two males copulated with the same female during the same reproductive cycle they could continue to compete after copulation through the action of their sperm. Parker referred to this as sperm competition – a term that had been introduced previously by Otto Winge in 1937 to describe the same phenomenon in guppies. However, Parker was the first fully to appreciate the evolutionary significance of sperm competition, defining it as the competition between ejaculates of different males for the fertilization of a female's eggs.[11]

Whereas Darwin's focus had been on the acquisition of partners, Parker's focus was on the acquisition of fertilizations. While Darwin assumed that gaining possession of a partner was sufficient, Parker pointed out that unless that male also fertilized the female's eggs, he would gain no genetic representation in the next generation. If sperm could compete for fertilizations within a female's reproductive tract, sexual selection would continue right up to the point of fertilization. By recognizing that sexual selection did not stop at insemination, and by combining this with what was then the new idea that what really matters in evolutionary terms is getting your genes into the next generation, Geoff Parker laid the foundations for the study of sperm competition. The essence of Parker's view was that if females copulated with more than one male, sexual selection would favour the males that fertilized most eggs. Males could win the fertilization contest by inseminating more sperm, by having faster-swimming sperm or by having sperm that disabled rival sperm. The way sperm win this contest is the basis for the study of sperm competition.

Parker's view of evolution, which had its roots in the work of George Williams and John Maynard Smith, among others,[12] was that selection operates on individuals rather than on populations or species as a whole. This individual-based view of evolution got its first major airing with the publication of E. O. Wilson's groundbreaking volume *Sociobiology*, published in 1975, and the next year was popularized by Richard Dawkins in *The Selfish Gene*. Implicit in Parker's ideas about sperm competition was the notion that far from being a co-operative venture between the sexes, reproduction was a selfishly motivated exercise, with each male and female out to maximize their benefits and minimize their costs. Sometimes the interests of each sex coincided, creating the illusion of co-operation, but most of the time each individual was out to get the best deal – even at the expense of his or her partner. This is sexual conflict: the battle of the sexes, where males and females are out to screw each other for the best, selfish genetic deal they can get. As we shall see, this unconsciously selfish attitude by each sex has been the driving force for many behavioural, physiological and anatomical aspects of reproduction.

Interestingly, for reasons that will become clear, Parker was initially less concerned with the battle *between* the sexes, than the battle *within* one sex: between males. Perhaps the most fundamental point Parker made about sperm competition was that it generates opposing selection pressures for males. On the one hand, sexual selection favours those males who successfully fertilize females who have previously been inseminated by other males. On the other, it simultaneously favours those males who prevent females *they* have inseminated from being fertilized by any other male. The benefits of securing fertilizations with additional females are considerable: more offspring bearing copies of that male's genes. But the costs of being cuckolded are also considerable. The reproductive success of a cuckolded male is depressed. And it can be reduced even further if, like ex-UK Prime Minister Harold Macmillan, cuckolded by Lord Boothby,[13] a male spends time and energy rearing offspring that are not his own – frittering away resources that could have been used to promote his own genes or secure reproductive success elsewhere. In the light of this it is ironic that Macmillan was famous for the phrase 'You've never had it so good.'

The outcome of the conflicting selection pressures created by sperm competition is the evolution of traits that promote male reproductive success. It is an evolutionary arms race. As soon as males evolve some character, such as a larger ejaculate, which increases their success, there is counter-selection on males better to protect their paternity, through, for example, behaviours that reduce the perceived likelihood of female infidelity. As soon as such behaviour has become effective, there is counter-selection on males to find another way to circumvent this.

Geoff Parker is indisputably the father of sperm competition. Multi-talented – he is also a jazz musician and breeder of champion chickens – Parker has been described recently as the 'professional's professional'. His ideas from the late 1960s and early 1970s provided the basis for all subsequent research in this field. Not only that; over the past thirty years he has continued to expand this theoretical base and erect a conceptual superstructure around which empirical tests of his ideas have been performed.[14] Layer by layer, year by year, the subtlety and complexities of post-copulatory male competition continue to be revealed.

Had Darwin Thought of Sperm Competition?

In almost all his voluminous writings Charles Darwin assumed that females were monogamous and copulated only with a single partner in each breeding attempt. Bob Smith of the University of Arizona, who edited the first compendium on sperm competition published in 1984, accused Darwin of delaying the study of sperm competition because he assumed females to be monogamous. Was Darwin really that naïve? Like his grandfather, Erasmus, Darwin was aware that sexual reproduction was the principal source of variation on which natural and sexual selection worked to cause evolution. Moreover, both Erasmus and Charles had read Spallanzani's *Essay on Animal*

Reproduction of 1769, in which the process of fertilization was first inferred. Given Darwin's recognition of the importance of sexual reproduction, it is hardly surprising that during his brainstorming years following the return of the *Beagle* he should have recorded in his notebooks many aspects of reproduction, including infidelity in chickens and multiple paternity in dogs.[15]

Later, Darwin had a protracted opportunity to think about sperm competition – in barnacles. Darwin's eight tedious years of barnacle dissection were enlivened by two discoveries. The first was that not all barnacles were hermaphrodite; some species had separate males and females. But, and this was the second revelation, in many species the males were minute with several living parasitically inside a single female. Stunned by his discoveries, Darwin wrote and told his friends.[16] To Charles Lyell, his geological mentor, he described the situation in one particular species:

> *... the other day I got the curious case of a unisexual, instead of a hermaphrodite, cirripede [barnacle], in which the female had the common cirripedal character, and in two of the valves of her shell had two little pockets, in each of which she kept a little husband; I do not know of any other case where a female invariably has two husbands ...*

In other barnacle species Darwin found as many as fourteen miniature males inside a single female. These findings were so extraordinary that much of his barnacle monograph is concerned with convincing his readers that these diminutive and rudimentary organisms really were males of the same species and not parasites. Darwin described them as mere bags of spermatozoa and in discussing their general biology made an implicit statement that both or all males inside a particular female could potentially fertilize her eggs. Subsequent barnacle studies have verified the accuracy of Darwin's observations and speculations, although it still remains to be shown the extent to which the sperm from the different males within a single female compete to fertilize her ova.

After completing his barnacle volumes, Darwin decided in 1855 to breed pigeons. As anyone who has kept them will know, pigeons provided Darwin with an excellent opportunity to observe courtship and reproductive behaviour at close range. And in a throwaway line in his rarely cited book on *Domestication*, Darwin provides one of the first ever references to extra-pair behaviour in male birds:[17] 'Pigeons ... can be easily mated for life, and, though kept with other pigeons, they rarely prove unfaithful to each other. Even when the male does break his marriage-vow, he does not permanently desert his mate.'

Perhaps the most explicit of all Darwin's statements regarding sperm competition appears in his volume *Sexual Selection and the Descent of Man*. The information originated from his cousin, William Darwin Fox, rector at Delamere in Cheshire. Fox kept a menagerie of farmyard animals, which over the years provided him with numerous observations and anecdotes for his cousin. Fox recounts that he had two types of goose: a gander and three female common geese and a pair of Chinese geese. The two types of goose had kept quite separate until one year the Chinese gander seduced a female common goose. When the eggs of this female hatched, it was clear from the appearance of the goslings that four of the eggs had been fathered by the common gander, the other eighteen by the Chinese male. So, as Darwin wrote: 'The Chinese gander seems to have prepotent charms over the common gander.' A clearer observation of sperm competition would be hard to find.

Darwin also knew about the potential for sperm competition in humans. When he wrote to his life-long colleague Joseph Hooker, who was on a botanical expedition in Bhutan, to tell him about the supplemental males in barnacles, Hooker wrote back acknowledging that Darwin's barnacle discoveries were wonderful. But, he said, they paled into insignificance in comparison with the polyandrous humans he was encountering, where 'a wife may have 10 husbands by law'. In their racy biography of Darwin, Adrian Desmond and James Moore interpret this as Hooker drawing a parallel between primitive people and primitive animals, both lacking the virtues 'innate to Gentlemen of the highest Victorian class'.

With so much evidence for sperm competition staring Darwin in the face, why didn't he pursue the idea within his framework of sexual selection? One possibility is that he simply never made the connection. In a riposte to my suggestion that he had falsely accused Darwin of delaying the development of sperm competition, Bob Smith proposed that Darwin failed to recognize the magnitude of its evolutionary implications.[18] However, given Darwin's perceptive ideas about most other areas of biology this seems unlikely. It is much more plausible that Darwin was inhibited. Victorian values made it unacceptable for someone in his position to discuss the nitty gritty of animal reproduction. It was admissible to discuss fertilization and illegitimacy in plants, as Darwin did, but the line between plant and animal sex was a rather clear one. Earlier, in the seventeenth century, sexual reproduction was discussed openly, and even in the late 1700s it was still acceptable and fashionable for Grandfather Erasmus to be a libertine. His monumental poems on the reproduction of plants were highly acclaimed by contemporary critics, although (or perhaps precisely because) they were little more than thinly veiled erotica. But by the time Charles was writing about sexual selection things had changed dramatically and Victorian prudery was well established. Thought to have arisen as a consequence of the Industrial Revolution and the rise of capitalism, the Victorian repression of sex was seen as a way of controlling the masses, preventing them from dissipating their energies and permitting them to do only what was essential to maintain the workforce. Whatever its cause, sexual repression made it difficult for Darwin to discuss animal reproduction. Darwin's thoughts were not repressed, however, and it was only in those writings he assumed would be read by the susceptible masses themselves that he avoided discussing animal sex. Because he doubted whether anyone other than broad-minded academics would ever read his barnacle volumes, which were rather technical, they are refreshingly frank when it comes to reproductive anatomy. Darwin describes with uninhibited enthusiasm the penis of one species being 'wonderfully developed', lying 'coiled up, like a great worm' and when 'fully extended, it must equal between eight or nine times the length of the animal'. Elsewhere in these volumes he discusses other aspects of barnacle *genitalia* and *spermatozoa*, terms that never saw the light of day in his more popular books.

In a sense this omission is hardly surprising since, as its title implies, *The Descent of Man and Selection in Relation to Sex* was ostensibly a book about Man, despite the fact that more than half the book is concerned with other animals. Had he so much as breathed a whisper about competing sperm, his readers would soon have put two and two together and made six. In fact, Darwin probably never stood a chance of discussing any sexual topic in much detail because his wife Emma and more particularly his hypochondriac daughter Henrietta acted as his censors. Darwin was sixty when he wrote *Descent*; he was slowing down and willingly handed over some of the responsibilities of correcting proofs and refining the text to his daughter. Henrietta has been described as both a closeted Victorian matriarch and a fussy moralist and it seems likely that she exerted a powerful force over what was and what wasn't morally acceptable in her father's books. As well as editing *Descent*, she gave Darwin's biography of Grandfather Erasmus the once-over, slashing out the bits she felt were unsuitable, including Charles's statement regarding his grandfather's 'ardent love of women'. Some idea of Henrietta's outlook on the world can be gauged from the fact that she initiated a campaign to eradicate the stinkhorn fungus, whose scientific name, *Phallus impudicus*, at once describes it and explains why she felt the sight of one might be a bad influence.[19]

Ironically, it was, I suspect, the ghost of Erasmus Darwin that provided the ultimate deterrent and prevented Charles from developing any ideas he might have had about sperm competition. Erasmus was obsessed by sex, both in theory and in practice. At a theoretical level he was well aware of the value of sexual reproduction in creating variation. On a more practical note, he prescribed it to his patients as a cure for hypochondria. But far worse from the family's point of view, Erasmus had fathered illegitimate children. The family had always known about this, but precisely at the time Charles was writing *Descent*, sixty years after his grandfather's death, rumours were circulating about the widow of a friend being an illegitimate granddaughter of Erasmus. Any discussion of sperm competition, however dispassionate, could very easily have backfired, and the subject was therefore left well alone.[20]

A Battle within a Battle of the Sexes

So far our focus has been on males and on sperm competition. This is deliberate. However, recall that I mentioned briefly that early in the history of sperm competition it was recognized that selection operated on each sex separately and how, in evolutionary terms, each sex was designed to maximize its own reproductive success. But – and this is a very significant 'but' – this battle between the sexes was seen as a very asymmetric one with active males and virtually passive females. Males fought for females; males displayed to females; males inseminated females, and their sperm then battled to fertilize a female's eggs. The most females were thought to do was to choose between different partners. This view, prevalent throughout the 1970s, appeared to be little changed from one proposed by Walter Heape in 1913:[21]

The Male and the Female individual may be compared in various ways with the spermatozoa and ovum. The Male is active and roaming, he hunts for his partner and is an expender of energy; the Female is passive, sedentary, one who waits for her partner and is a conserver of energy. To the Male it is the sexual act which is of moment, while it is the consequence thereof which profoundly affects the Female.

This androcentric view of reproduction goes back to Aristotle, and despite Darwin's ingenious idea about female choice, sexual selection was still a male-dominated process. This gender bias, which stemmed from a mixture of unconscious sexism and biological ignorance, persisted and formed an important part of Parker's and Trivers's formulation of sperm competition. However, Parker did at least provide a biological reason for his male-orientated vision. It was this: a male who copulates with, and fertilizes, as many females as possible (including some previously inseminated by other males) will produce more descendants than one who does not do so. For a female, however, copulating with more than one partner is unlikely to increase the number of offspring she produces; the best she can hope for is an increase in their quality. Parker's view was based on the idea that evolution is a game of numbers – what counts is how many genetic representatives an individual has in subsequent generations – and because quantity is more important than quality, sexual selection operates more intensively on males than on females.[22]

The basis for this argument was female passivity, the idea that females simply acquiesced. This view may have been an accident of the fact that the first sperm competition studies were conducted on insects, animals so far removed from ourselves that it was difficult to tell what their motivation was. It was not until the 1980s when researchers started to look at sperm competition in vertebrates, and in birds in particular, that it was realized that, far from being passive, females often actively sought multiple male partners. With this realization there was a major shift in the emphasis of research. If females choose their partners, and if they actively choose to copulate with several males, might they not also choose between the sperm of different males? Just as sperm competition had been a continuation of pre-copulatory male–male competition into a post-copulatory context, so sperm choice was a continuation of a female's pre-copulatory choice of a sexual partner. Moreover, as soon as it was recognized that females might choose between the sperm of different males, it was clear that there would be conflict between the sexes. Whenever one sex, let's say it is the male, constrains the reproductive success of the other, selection immediately favours an adaptation in females to overcome the constraint. The result is sexual conflict – the battle of the sexes – and an escalating arms race of adaptation and counter-adaptation between males and females. Put simply: sperm competition + sperm choice = sexual conflict.

It is probably no accident that the increased interest in female aspects of behavioural ecology, including sexual selection, coincided with the continuing expansion of the feminist movement. Bob Smith told me that when the book on sperm competition he edited was published in 1984 his chapter on sperm competition in humans triggered off an angry response, particularly among militant feminists. The feminist movement in North America in the 1970s was epitomized by a popular bumper sticker, attributed to the journalist Gloria Steinem, which read: 'A woman without a man is like a fish without a bicycle.' For women who didn't need a man at all, the idea of having several to induce or facilitate sperm competition was anathema.

This story also reflects how female academics interested in sexual selection on opposite sides of the Atlantic differed in their approach to the gender bias in behavioural ecology. Without doubt, the most radical feminists were (and still are) North Americans. By contrast, those in Europe adopted a much more subtle, and ultimately probably a more persuasive, strategy in their efforts to change the way (male) biologists thought about reproduction. Moreover, it is important to recognize that amending the gender bias has not been an entirely female prerogative; several male behavioural ecologists have actively promoted the female perspective. It is unfortunate, therefore, that these include several who continue to be, until recently at least, targets of North American feminist criticism.[23]

In terms of recognizing the importance of the female perspective we have come a long way in the thirty years since sperm competition was first identified as a component of sexual selection. Some feminists might still argue that we have not come far enough, but the majority of behavioural ecologists I talk to think that the balance is now reasonably even. Despite this substantial shift in outlook, research effort over the past few years has done little to change the view that sperm choice and the interactions between sperm competition and sperm choice are often extremely subtle and remarkably difficult to elucidate. As we shall see later, this is exemplified by the difficulty behavioural ecologists have had in identifying both the mechanisms by which females control the paternity of their offspring (chapter 6), and the adaptive significance of polyandry itself (chapter 7).

Reproducing Controversy

Like many new areas of science, the combined field of sperm competition and sperm choice stimulated a huge amount of innovative research. Indeed, it is recognized that one of the main ways in which science makes progress is through a set of new ideas resulting in a fundamental change in outlook. This is referred to as a paradigm shift[24] and for those involved it is one of the most exhilarating aspects of science. Paradigm shifts are characterized by intense excitement and high productivity, followed, as things settle down, by a period of what is referred to as 'normal science'. Depending on the magnitude of the shift and the type of science involved, the initial phase of enthusiasm and productivity may last months or years. Paradigm shifts might be the engine that drives scientific discovery, but at the same time they are open to exploitation via what I call 'the bandwagon effect'. The wave of enthusiasm for a fashionable new area of research creates an opportunity, utilized by some researchers, for the publication of hastily conceived and poorly executed research, thereby allowing them to jump aboard the bandwagon.[25] This process is facilitated by two factors. First, new areas of research have, by definition, few experts capable of critically assessing the work of others. Second, the editors of scientific journals, keen to promote their journal by publishing results from topical areas of research, tend to be less critical than they are at other times. Paradigm shifts have probably always been accompanied by some sloppy science, but it is my belief that it has become much more obvious as competition for research funds and faculty positions has increased.

Sperm competition and sperm choice have not escaped the bandwagon effect, and one particular area to have suffered in this way has been the study of human reproduction. Perhaps we should not be surprised that it has been the study of human sperm competition that has been problematical. We are obsessed with our own biology and this renders anything both biologically human and novel particularly vulnerable to exploitation and uncritical media publicity. The new field of evolutionary psychology, a discipline that seeks adaptive explanations for such things as jealousy, infidelity and status enhancement, provides a clear example.[26] Similarly, you have only to think about the intense interest occasioned by new hominid fossil discoveries – nothing could be less sexual, but, as media coverage indicates, this is sexy research. This in turn should make us question why researchers opt for particular areas of science. Would archaeologist and anthropologist Louis Leakey have attained the same international acclaim and status had he studied fossil fish? I think not. Working with human origins, both archaeological and biological, greatly increases a researcher's media profile, and all that goes with it. So perhaps the inclination to explore and exploit such fields is motivated, in part at least, by status. Leakey's reputation made him immensely attractive to women, and after meeting him for the first time (in 1959) primatologist Irven DeVore said to his wife, 'He must be one of the ugliest men I've ever met. What do you think?' Her reply was, 'Are you kidding? That's the sexiest man I've ever laid my eyes on.' Others felt the same and there was rumoured to be a trail of Leakey's progeny all the way from his home to the Ulduvai Gorge.[27] Similar stories exist for almost all successful men.

The first person to view human sperm competition in evolutionary terms was Bob Smith.[28] His chapter, the final one in the compendium on sperm competition he edited, was a monumental achievement and the culmination of eighteen months' research and writing. To me at least this was a scholarly, objective, no-nonsense account. It was also extremely stimulating, pointing out the new insights an evolutionary approach to human reproductive biology could provide. Bob's review was totally hands-off. It was based entirely on the information he could glean from the literature – other people's studies. But it stimulated Robin Baker, an evolutionary biologist then at the University of Manchester, to initiate a hands-on project on human sperm competition – and this was when the trouble started. Baker and his collaborator Mark Bellis plunged in where no one else had ever dared to probe – combining the *risqué* notion of infidelity and evolutionary ideas. They joined a long line of illustrious researchers interested in human sexuality: Havelock Ellis in the early 1900s, Alfred Kinsey in the 1940s and 1950s and William Masters and Virginia Johnson in the liberated 1960s and 1970s. In a critical review of the efforts of these early sexual pioneers, Paul Robinson speculated, in 1976, about future studies of human sexuality.[29] The two issues he felt would become the focus of further research were homosexuality and – anticipating Baker and Bellis – extra-marital sex. Regarding the latter, Robinson wrote, 'What seems to be called for is a theorist of stature who will contemplate the phenomenon sympathetically, but without giving way to uncritical effusings.'

Robin Baker, now retired, was an innovative biologist. He had taught me when I was an undergraduate at Newcastle, telling my third-year class about Bob Trivers, about individual selection and about Geoff Parker's sperm competition studies on dungflies – before they were even published. For me, Robin's lectures were electric and the most memorable of my entire undergraduate career. As he explained the basics of individual selection and the evolutionary conflicts created by sperm competition it was like a light going on. At that time Robin's own research was on insect migration; he simply told us about sperm competition. But Robin loved controversy: he had novel ideas and sold them hard – but at a cost. He told us how he had been scorned by other insect biologists. In response, he produced a doorstop volume designed to halt his critics in their tracks – by its bulk if not by its arguments. And Robin has courted controversy ever since. Following on from his studies of insect migration, when Robin moved to Manchester University in 1972, he started a study of human navigation – testing the idea that inside our heads we all harbour a fundamental magnetic sense. Generations of undergraduates were blindfolded and bussed around Manchester. On their eventual release they were asked to point in the direction of home. Baker alleged that, despite their nausea and vomiting, they could. But the statistics were complicated and others working in navigation research found the results less than convincing. This was high-profile stuff. The media loved the idea that somewhere in our skull there were little magnets that realigned themselves to allow us to find our way home. Sceptics around the world repeated Baker's study, but none was convinced that he had replicated Baker's results. Undeterred, Baker re-analysed the information from these other studies, combined them and claimed convincing evidence for his ideas.[30] Having resolved this particular problem and now appreciating the wealth of data that could be generated from human subjects, especially enthusiastic undergraduates, Baker turned next to sex and human sperm competition.

Initially at least, those biologists working on sperm competition in other animals were optimistic, and admired Robin's audacity. At the International Society for Behavioural Ecology Congress in Princeton in 1992 Baker and Bellis were billed to give consecutive talks in the same session. The auditorium was packed. Wearing skimpy shorts and a shirt open to his navel, Baker kicked off, and put on the kind of show his audience had come to expect. Titillating; feeding our irrepressible curiosity about our own reproductive performance, Robin Baker, and later Mark Bellis, recounted how their nationwide survey had revealed high levels of polyandry, of how the sperm from competing men engaged in warfare – killing each other inside the female reproductive tract – and of how females regulated this battle by using their orgasm to control the uptake of sperm and ultimately the paternity of their offspring.[31]

Simply by standing up there and telling us things few had ever dared discuss in public, Baker's status was enhanced. I couldn't help being impressed by his immoderate ideas and his ability to bind an audience, but I was also deeply sceptical. My scepticism stemmed from an incident that had occurred two years previously, at a much smaller meeting I had organized in Sheffield. At that stage, I had faith in Baker's and Bellis's work – but this was the watershed. The meeting was arranged to encompass a wide range of interests in sperm biology. I was concerned that as behavioural ecologists dabbling in sperm we were missing out on the huge amount of information and knowledge possessed by reproductive physiologists and andrologists – people we rarely came across, let alone exchanged ideas with. Accordingly I had invited a number of people from rather different disciplines, including two high-ranking reproductive physiologists – Harry Moore, who had recently taken a position in Sheffield, and Mike Bedford from Cornell University Medical College, New York.

Robin Baker gave a confident performance, but one that was scientifically difficult to defend. He proposed that if differences in relative testis size between species could predict their involvement in sperm competition (see chapter 3), the testis size of individual men could predict their success in sperm competition. In conducting this study Baker had persuaded 14 of his colleagues at Manchester to measure the size of their left testicle with a pair of callipers. He then asked 20 female colleagues of the men (who apparently had no knowledge of their testis size) to rank these men according to the likelihood that they would engage in an extra-pair copulation if given the chance. Baker presented the results – and, just as he had predicted, there was a significant, positive correlation between testis size and the perceived likelihood of a male engaging in an extra-pair copulation. The audience response was mixed: sniggering incredulity at Baker's audacity and, from some quarters, downright indignation.

Baker's approach seemed naïve. In reporting testis size he had not taken into account the height of the men; that is, he had not expressed testis size in relation to male height. In other animal groups it is quite clear that bigger individuals have bigger brains, bigger hearts and bigger testes. Were all the men in Baker's sample Caucasian? Racial differences in testis size are substantial.[32] Indeed, a whole range of confounding variables could account for the positive relationship, none of which was considered or dealt with. At question time the mood turned distinctly sour as the reproductive physiologists challenged Baker's rationale. As the session ended, an indignant physiologist hounded me back to my office. What did I think I was doing bringing him across the Atlantic to witness this caricature of science?

Worse was to come. Soon after Baker's and Bellis's results were first published the media started to take notice of their work. Two years after the Princeton meeting behavioural ecologists gathered once again, this time at Nottingham. By coincidence Desmond Morris's television series *The Human Animal* was running, and during the meeting the programme on human reproduction was broadcast. To the UK's television-watching public, Desmond Morris, made famous by *The Naked Ape*, continues to be one of the foremost authorities on matters zoological. The basis of this programme was Baker's and Bellis's work and it started by illustrating one of their most novel ideas: the concept of killer sperm.

The ejaculates of all animals contain a range of sperm types – sperm that vary in their morphology: big heads, little heads, straight tails, curly tails, and so on. Conventional wisdom has it that the testes find sperm tricky to make, and so errors occur: the morphological variations are the duffers, the seconds, the rejects. Baker and Bellis turned that idea on its head.[33] These weren't mistakes, they said. The different types of sperm had different roles to play. They had evolved to be different. This was behavioural ecology speak – finding an adaptive explanation for something previously thought to be evolutionarily neutral or even maladaptive. The role for certain sperm types was, Baker said, to seek and destroy the sperm of rival males.

The opening image of Morris's television programme was of a human ovum with a shimmering halo of sperm. Having introduced Baker's and Bellis's concept of killer sperm in the first few minutes, Morris then went on to state that, 'New research suggests that a man can unconsciously control the number of killer sperm in his ejaculates and that would depend on whether or not he believed he was the first or second male to mate with a particular woman ... Here's one in action.' At this point we saw a human sperm bumping against what looked like, according to the andrologists I have asked, a sperm that was already dead. Morris continued, 'This is chemical warfare filmed for the first time ... and when the two sperm separate one is left dead.' Heady stuff, and exactly the sort of ooh–ahh science the media and its susceptible audience love. But wait a minute: even if this really was one sperm killing another, since these sperm were not labelled in any way, how could Baker, or anyone else, know that these sperm were those of different males?

Baker and Bellis were on a rollercoaster. The idea that the different types of sperm within a human ejaculate each had a specific role in sperm competition was very sexy. Initially there were just 'killer sperm', and these formed the basis for what Baker and Bellis called their 'kamikaze sperm hypothesis'. With time they also identified other sperm types, including 'blockers' and 'egg-getters'. Killer sperm used their acrosome, the package of enzymes at their tip, to destroy the sperm of other men. Blockers were those with coiled tails designed to impede the progress of any rival's sperm. And the egg-getters were the tiny proportion of sperm with large heads whose job was to fertilize.

Where was the evidence for this? Evidence of a sort was discussed at scientific meetings but not in a form that allowed anyone to scrutinize it. Conference presentations typically focus on results rather than methodology, since we are usually happy to assume that the techniques employed are appropriate. But here the methodological aspect was crucial to allowing others to assess the validity of these bold ideas. Subsequently, Baker and Bellis did present some information in their book, *Human Sperm Competition*, claiming that, after mixing the sperm from two different men, they could see sperm stuck together in the process of killing each other. They also saw an increase in the proportion of acrosome-reacted sperm (indicating that they had used their acrosome in the killing process), and they also saw casualties – many more dead sperm in mixtures than in single-sperm samples. Taken at face value it reads fairly convincingly, especially if you have no additional information. But there were other studies that had mixed sperm from different males, albeit not from humans but from domestic mammals and birds, and none of them had reported anything resembling warfare. Subsequently Harry Moore, working in collaboration with colleagues at a local fertility clinic, reran Baker's and Bellis's experiment, but at a much more sophisticated level. They differentially labelled the sperm from the different men to see whether the agglutination Baker and Bellis had reported was between sperm from different males, as it would have to be. But no. Whatever agglutination there was was random with respect to male. Moreover, mixing sperm from different males caused no induction of the acrosome reaction and no increase in sperm mortality.[34]

Other parts of the specialist sperm story also started to disintegrate.[35] In a review of their book, Roger Short pointed out that the sperm Baker and Bellis designated as 'egg-getters' were about as likely to be egg-getters as they were to contain little men. Twenty years previously Short and three colleagues had published a paper in the journal *Nature* pointing out that these large-headed sperm were production errors carrying twice the normal chromosome complement and as a consequence were quite incapable of producing a normal embryo. Somehow Baker and Bellis had overlooked this publication.[36]

Baker's and Bellis's claims regarding human sperm competition were momentous and caught the public imagination almost to same extent as Stephen Hawking's *A Brief History of Time*. Why? Results from studies of sperm competition in other animals – which I describe later – equally exciting, equally bizarre and far better substantiated, hardly ever make the press or television. Why are we obsessed with ourselves and with our own reproduction? In part at least, what Baker and Bellis reported was so extraordinary, one couldn't help but be amazed: one had only to imagine sperm warfare going on inside female bodies . . . Another part of the fascination might have been the idea that knowledge is power: knowing something of how sperm competition occurs might make us better, or possibly more successful, lovers. Baker's and Bellis's initial credibility stemmed from the fact that their studies were based on a firm body of theory provided by other biologists and on the fact that sperm competition and sperm choice, as will become apparent, occur in myriad different forms throughout the entire animal kingdom. Indeed, responding to his critics, Baker has said that had he obtained his results with any other species they would have been accepted without question.

The reproductive adaptations proposed by Baker and Bellis appear to be more extreme than is merited by the level of sperm competition that seems likely to have occurred in our ancestors (chapter 2). It is not surprising then that, so far, many of Baker's and Bellis's results have failed to stand up to scrutiny. Robin Baker and Mark Bellis may see themselves as pioneers in the study of human sexuality, but I fear that Paul Robinson, wary of 'uncritical effusings', would have been disappointed by their efforts.[37] The view of human sperm competition they have perpetuated is little more than a sexual fantasy – phallus in wonderland – and I have exorcized it here in order to leave the way clear for a more veracious but no less astonishing account.

The Main Questions

Once the evolutionary implications of sperm competition had been spelt out by Geoff Parker in 1970 the potential for sperm competition to explain many biological phenomena was obvious. Following a seven-to-ten-year period of gestation typical of some research areas, the subject eventually exploded in an exponential increase in interest. Several factors made sperm competition popular. First, it concerned a topic close to our hearts: reproduction. Second, because it dealt directly with variation in reproductive success its evolutionary significance was much more immediate than that of other behaviours, such as foraging. And third, it opened up an entirely new field of research with the potential to explain things that had previously been inexplicable.

There are three main questions in the study of sperm competition and sperm choice. The first is why an individual should copulate with more than one partner. The second is concerned with the consequences of polyandry: what happens when a female copulates with more than one male; specifically, what determines which male fertilizes her eggs? The third question focuses directly on the battle of the sexes and asks how sexual conflicts – the inevitable consequence of sperm competition and sperm choice – are resolved.

The first question is typical of the type asked by behavioural ecologists and addresses the evolutionary or functional significance of copulating with several partners. It is the same as asking why copulating with several partners increases male and female reproductive success. This is a 'why' question: *why* has the tendency to copulate with several partners evolved? The second question is a 'how' question and is concerned with the underlying physiological mechanisms: *how* do copulations translate into fertilizations? The third question is also concerned with evolution and how each sex generates selection pressures on the other, an area of research referred to as antagonistic co-evolution.[38]

All three questions are equally important in helping us understand why particular reproductive structures, physiologies or behaviours have evolved. During the brief history of sperm competition and sperm choice most effort has gone into addressing the 'why' question. As in all branches of science, big questions like this can usually be broken down into lots of smaller questions, and the smaller 'why' questions have been concerned with adaptations to sperm competition. Why do males inseminate a particular number of sperm? Why do males remain close to females before and after copulating with them? Why do the females of some species initiate so many copulations with their partner? Why do females seek extra-pair copulations? More recently, researchers have also started to wonder about the 'how' question: how do physiological processes, including sperm choice, transform copulations into offspring? Researchers also want to understand what happens to the sperm of different males inside a female's reproductive tract and they want to know about the relative importance of sperm competition and sperm choice in determining which male fertilizes a female's eggs.

The battle of the sexes is an ancient concept. But the idea that the battle of the sexes takes place between sperm and eggs is very recent. The notion that conflict exists between males and females over which sperm fertilize a particular egg has revolutionized our view of reproduction and sexual relationships. Who are the winners and who are the losers in this battle? The answer is difficult to predict because although selection may operate more intensively on males than on females, since the competition between sperm is played out inside the female's body, it may be relatively easy for females to manipulate sperm. Sexual conflict challenges long-cherished assumptions about male and female roles in sex. It is also responsible for the staggering proliferation of reproductive structure and function across the animal kingdom – reflecting the ongoing battle between the sexes for reproductive control. Only by understanding the evolutionary basis for the diversity of strategies and counter-strategies in non-human animals can we hope to shed any light on our own sexuality.

In the following chapters we shall describe both the saga of the sperm getting to the egg and the history of sperm competition and sperm choice. We start, in the next chapter, by focusing on the adaptive significance of sperm competition from the male perspective. Chapters 3, 4 and 5 then consider the machinery and mechanics of reproduction, discussing in turn: genitalia, sex cells, and the processes of sperm transfer and fertilization. These chapters are important since some knowledge of the anatomy and function of the reproductive system of each sex is crucial to understanding how sperm and eggs interact. Moreover, the history of sperm competition and sperm choice is a much neglected but vital part of the development of our understanding of sexual reproduction. Primed with this information, chapter 6 explains the male and female mechanisms that determine which of several males fertilizes a female's eggs. Finally, in chapter 7, we return to the issue of who benefits and look at what females gain from copulating with several partners.

2 Paternity and Protection

Investigations into paternity are forbidden.
NAPOLEON (Article 340)

Charlie Chaplin's penchant for teenage girls and young women is well known. One of his numerous affairs backfired in June 1943 when Joan Barry announced to the press that Chaplin was the father of her unborn child. This was more than a little awkward since by this time Chaplin's affair with Barry had finished and he was engaged to someone else. However, it was the FBI, not Barry, who initiated proceedings against Chaplin. The FBI was anxious to find an excuse to harass Chaplin because of his alleged Communist sympathies and used this paternity suit to pursue him over several years. Once the child was born a comparison of its blood group (type B) with that of Chaplin's (type O) and Barry's (type A) proved beyond all doubt that Chaplin was *not* the father. Thinking himself vindicated, Chaplin was devastated to find himself a few months later subject to another trial over who would support the child. The prosecuting attorney, playing on Chaplin's reputation, declared that there was only one way to stop his 'lecherous conduct' and persuaded the jury to find him guilty, forcing him to pay child support anyway.[1]

The issue of paternity is at the core of much of men's behaviour – and for good evolutionary reasons. In our primeval past men who invested in children which were not their own would, on average, have left fewer descendants than those who reared only their own genetic offspring. As a consequence men were, and continue to be, preoccupied with paternity and this has shaped not only many male behaviours but, perhaps surprisingly, some female behaviours as well. The most obvious way in which men's preoccupation with paternity manifests itself is in jealousy – watching a partner and keeping her away from potential competitors. At its most extreme, concern over paternity is responsible for a whole plethora of unpleasant acts by men against women: purdah, claustration, foot-binding, chastity belts, clitoridectomy, and downright murder.[2]

Paternity is such a fundamental issue, both for human males and for anyone with a more general interest in sperm competition, that we shall begin this chapter by ignoring Napoleon's edict and look at the ways in which paternity has been investigated. We start with the least reliable techniques and finish with the spectacularly sophisticated molecular methods currently available. We then examine the results of recent paternity studies which allow researchers for the first time to measure male reproductive success accurately and hence to establish whether males benefit from copulating with females even when they have been inseminated by other males. We then consider what the results of paternity studies tell us about polyandry and sperm competition. Finally, we look at the way males cope with the threat of cuckoldry – their behavioural adaptations to sperm competition.

Measuring Paternity

Because they spent so much time away from their wives, cuckoldry was an occupational hazard for Mediterranean shepherds and it was apparently not uncommon for each child in a family to be fathered by different a man.[3] Whether the shepherds themselves were aware that they had been cuckolded is not recorded. The only thing they could go on was the similarity – or lack of it – between themselves and their offspring. Even if they had their suspicions they can never have been absolutely certain whether a particular child was theirs because the extent to which fathers and true offspring resemble each other is so variable.

Under the right circumstances, however, some forms of similarity between fathers and offspring, so-called genetic markers, can provide unambiguous evidence for extra-pair paternity. Indeed, it was through the unintentional use of genetic markers – probably differences in plumage colour – that Aristotle was able to report mixed paternity in broods of the domestic fowl.[4] Occasionally genetic markers can also be used to assign paternity in humans. Because we now know, for example, that the genes for ginger hair are recessive, a couple who are both redheads will normally produce only ginger-haired offspring. The birth of a dark-haired child therefore would suggest that the woman had been involved in an extra-marital affair with a dark-haired man. Now imagine a ginger-haired couple who produce two offspring: one with black hair, the other with ginger hair. We can be sure that the husband was not the biological father of the black-haired child, but we would naturally assume that he *was* the genetic father of the ginger-haired child. However, we cannot be absolutely certain of this because there is an outside chance that this offspring was fathered by a different male (ginger or not) carrying the ginger gene. The likelihood of this depends upon the proportion of ginger-genes in the population as a whole – the fewer there are, the less likely it is that the ginger-haired child was the result of an extra-pair copulation.

It is not too difficult to see how if one restricted the set of mating partners one could use genetic markers to assign paternity in non-human animals. Robert Nabours, a research scientist at the Kansas State Agricultural College in the 1920s, did this with grasshoppers and conducted what amounts to the first systematic study of sperm competition. In his own words, his experiments were motivated by the 'mere curiosity to determine how many males might take part in the parentage of the progeny of given females'. He exploited the fact that female grasshoppers will copulate in rapid succession with several males and that the grasshoppers existed in different colour forms, enabling him to assign paternity to individual males. Nabours's experiments showed not only that a clutch of eggs could have several fathers, but that the last male to inseminate a female usually fertilized most of her eggs.[5]

For animal species without genetic markers researchers needed another method to establish paternity. One solution was the use of so-called sterile males. This technique, first used in 1939 in a study of silk moths,[6] works like this. Take some male insects and place them – carefully – in front of a radiation source. With doses that would send us to an early grave, the only effect on male insects is to scupper their sperm. But not completely. The trick is to deliver a dose of radiation sufficiently small that the sperm are still capable of fertilization, but sufficiently large to damage the sperm's chromosomes so that, once fertilization has occurred, the egg is incapable of proper development. If done correctly, the result is a batch of fertilized eggs, all of which fail to turn into embryos. The term 'sterile male' is therefore something of a misnomer. An irradiated male is perfectly capable of copulating and his sperm are able to fertilize, but the essential feature is that eggs fertilized by the sperm of an irradiated male do not develop. The essence of this method, then, is to arrange for a female to copulate with both a normal male and an irradiated male: those eggs that develop were fertilized by the normal male; those that do not were fertilized by the sterile male. Although several early researchers used this approach, it was Geoff Parker in the 1970s who first developed and exploited the potential of this method in the study of sperm competition.[7] Across a range of insect species the result was that the last male to copulate with a female fertilized the majority of her eggs. This strong order effect, referred to as 'last male sperm precedence', confirmed that copulating with an already inseminated female was highly adaptive for males.

Charlie Chaplin was lucky (or should have been) that his exclusion as the father of Joan Barry's child was so clear cut. Almost any other result would have implicated him as the father, even though one thing a blood-group test cannot do is unequivocally assign paternity to particular man. If Joan Barry's child had been blood group O, Chaplin *could* have been the father but, given that this is the commonest blood type, so could many other men and this was the major limitation of this and other contemporary paternity methods. Genetic markers and sterile males satisfied sperm competition researchers as a way of assigning paternity only as long as they were happy to confine their experiments to the laboratory where they could arrange by which males a female was inseminated. But as soon as they started to ask whether litters of mammals or broods of birds in the wild might have multiple fathers, the shortcomings of these two techniques were all too apparent. The answer was – eventually – molecular methods.

The first molecular parentage studies were made in the early 1970s using what now seems to be the relatively primitive technology of allozyme analysis. This technique relied on a process called gel electrophoresis to detect differences between individuals in the proteins comprising the enzymes contained in their blood or other body tissues. It was similar to the genetic marker method of paternity analysis in that allozymes could say only who was *not* the father, rather than assign paternity directly. The allozyme method relied on there being considerable variation, or polymorphism, between individuals in their enzymes. This variation existed in certain animal groups, making allozyme analysis a useful tool for detecting multiple paternity, and some of the earliest results, published in the 1970s and 1980s, revealed extensive mixed paternity in animals as different as deer, mice and lobsters.[8] For birds, however, the allozyme approach to parentage was assumed to be a non-starter because it had long been recognized that birds were notably lacking in the required enzyme variation. There was one striking exception, however. Through a combination of skill, perseverance and possibly a lucky choice of study species, David Westneat, now at the University of Kentucky, managed in the mid-1980s to wring rich results from the allozymes of a bird called the indigo bunting.[9] For a species assumed to be basically monogamous, his study exposed staggeringly high levels of extra-pair paternity – no less than 35 per cent of all offspring were fathered by extra-pair males.

The real revolution in molecular paternity studies took place in 1985. In that year Alec Jeffreys of Leicester University, who was studying the human myoglobin gene, found a way of visualizing minute differences in the DNA of different individuals. This method also uses electrophoresis, but this time of pieces of DNA itself (rather than the proteins for which it codes). Using certain sorts of DNA, electrophoresis produces a unique pattern of bands on a gel – a DNA fingerprint – for each individual.[10] The beauty of this method was that for the first time it permitted paternity to be assigned to particular individuals. It allowed forensic scientists to identify rapists unequivocally by matching semen from a woman to a blood sample from a man. In the study of behavioural ecology it allowed biologists to look at relatedness between individuals in more detail than they ever previously imagined and to establish whether multiple partners for a female meant multiple fathers.

By a fortunate coincidence, Terry Burke was also at Leicester, working on relatedness in sparrows, just at the time when Jeffreys made his discovery. Burke and his student Mike Bruford were quick to exploit this unique opportunity and, together with another group at Nottingham University, they found that what worked well for humans also worked well for sparrows. These two research groups were the first to use the new technology to look at parentage in birds and their papers were published back to back in the same issue of *Nature* in 1987.[11] These results generated a cascade of similar studies – albeit a slow-motion cascade. The only negative aspect of this monumental development was the difficulty of producing DNA fingerprints. The process was, and still is, relatively slow and tedious, even though the methodology has become increasingly refined and streamlined. The results spawned from this new technology have changed our view of life.

Paternity Revealed

The single most striking result from the slow accumulation of paternity studies has been the near elimination of the idea of male and female sexual monogamy. From organisms as different as snails, honey bees, mites, spiders, fish, frogs, lizards, snakes, birds and mammals, research has verified behavioural observations of females' polyandry by showing that multiple paternity is widespread.

Almost from the start it looked as though behavioural ecologists studying birds had a monopoly on molecular techniques for assigning paternity. This bias is due to the fact that birds were so much more tractable study organisms than most other animals. At the simplest level, a study of parentage in birds requires only that one obtains DNA from putative mum and dad, and from their offspring, which are usually packaged conveniently together in a nest. In marked contrast, the polygynous mating arrangements of most mammals mean that candidate males are much harder to identify and investigators have to sample several putative fathers. Add to this the difficulty of identifying and watching most mammals and the fact that young are often not reared in a nest, and you can see the problems.

Like many other people across North America, Gene Morton, a behavioural ecologist based at the Smithsonian Institution, Washington, DC, has a special bird house in his backyard for purple martins to breed in. People have been providing multi-apartment homes for purple martins for years and have enjoyed their comings and goings, assuming that they all live in perfect harmony. Morton decided to mark and watch his martins and to look at the paternity of their offspring. His findings were remarkable. The first birds to arrive in the spring were adult males. They laid claim to the bird house and sang to attract a female. Other males were aggressively chased off. Once the house owner had secured a partner and she was incubating her eggs, the male underwent a change of personality. Now, other males were welcomed rather than rejected. In due course these males, which were all young, first-time breeders, occupied rooms within the house and attracted their own partners. However, when Morton scored the paternity of the chicks in each nest he found that the house-owning male not only fathered an average of 96 per cent of offspring in his own nest; he also fathered over half of all the offspring in the nests of other males. After aggressively ensuring paternity with his partner, the house-owning male then appeared deliberately to attract other males. Breeding sites are in short supply and these young males are desperate for an opportunity to breed, but having lured them into his house, the older male then cuckolded them.[12]

In the years following the discovery of DNA fingerprinting more and more biologists decided to measure the frequency of extra-pair offspring in apparently monogamous birds and information is now available for over one hundred species, and the number of studies continues to grow. Two main results emerge. First, extra-pair paternity is widespread, occurring in about 70 per cent of the species examined. Second, the variation between different species is immense, with anything from zero to 76 per cent of offspring fathered by extra-pair copulations.[13]

Among the more sexually monogamous species is the fulmar. Fiona Hunter, then at Sheffield University, conducted a monumental study of this mini-albatross on Fair Isle off the north coast of Scotland. Taking four-hour shifts with a (dedicated) friend she watched a colony of 91 pairs of birds continuously for 18 hours every day for eight weeks, recording every single copulation – legitimate and illegitimate – and later collected blood samples from the birds to extract DNA and conduct the fingerprinting. Despite the fact that no fewer than 16 per cent of females were involved in extra-pair copulations, there was not a single case of extra-pair paternity.[14] At the time Fiona Hunter was disappointed by this apparently negative result, but it raised a number of interesting questions, and suggested that a male fulmar might have had more control over fertilization than his unfaithful partner might have wished. It was only later, as paternity studies of more species were published, that it became apparent that the fulmar investigation was a forerunner of a number of non-passerine bird studies in which extra-pair paternity was infrequent.

The zebra finches I studied in Australia had a similar level of extra-pair activity, but only a marginally higher rate of extra-pair paternity, than the fulmar: only 2 per cent of 91 zebra finch chicks were fathered by extra-pair males. Higher levels of extra-pair paternity occurred in other species. David Westneat used DNA fingerprinting to confirm his 35 per cent estimate for the indigo bunting, but this record was broken by the reed bunting, with 55 per cent extra-pair offspring. The most extreme case occurs in Australia where no less than 76 per cent of superb fairy wren offspring were fathered by extra-pair copulations.[15]

These results raise two questions: (1) what accounts for the variation in the incidence of illegitimate offspring in different species? And (2) does extra-pair paternity increase a male's reproductive success?

The most obvious answer to the first question is that the variation in extra-pair paternity can be explained by ecological factors. The distribution of food in both space and time determines whether a particular species is social or solitary, and it is possible that this might in its turn determine the opportunities for extra-pair sex. In particular, one might imagine that breeding socially in a colony or in an apartment block might increase the opportunity for extra-pair sex. But contrary to expectation, behavioural ecologists have failed to find any consistent relationship between the degree of sociality and extra-pair paternity. Indeed, they have failed to identify *any* ecological factor that accounts for the variation in extra-pair paternity. The answer to our first question is therefore that we simply do not know.

As far as the second question is concerned, we do have an answer. You might think it intuitively obvious that extra-pair copulations increase male success, but extra-pair fertilization is a form of parasitism and cuckolders succeed only at the expense of those they cuckold. It is possible that the benefit a male accrues from extra-pair fertilizations is offset by losses of paternity with his own female. In fact this is not the case because, as several studies have shown, there are clear winners and losers reflecting differences in male quality. Those males proficient at securing extra-pair fertilizations are also less likely to be cuckolded.

Anders Pape Møller, then at the University of Aarhus in Denmark, was the first to demonstrate this.[16] He showed that female barn swallows prefer males with long outermost tail feathers as both social and sexual partners. Moreover, by a clever experiment in which he increased the length of a male's tail, a male's attractiveness and reproductive success – particularly in terms of extra-pair offspring – were also enhanced. This, and similar studies, have shown that extra-pair copulations increase the reproductive success of attractive males and decrease it for less attractive individuals.[17] In other words, the effect of extra-pair copulations is to increase the *variance* in reproductive success among males in a population – some males father many offspring while others father few or none. Variation between individuals is the raw material of selection and is what drives evolution, so this result demonstrates the importance of sperm competition as a component of sexual selection.

Male and female swallows differ very little in their external appearance. The only detectable difference is the fact that the male's tail streamers are longer than those of the female. Since the length of a male swallow's tail streamer reflects his reproductive success, it seems likely that the sex difference in tail length among swallows is a result of sexual selection. In other species male and female birds differ more dramatically in their appearance and Møller and I found that across a range of different species the greater the sexual dimorphism the greater the extent of extra-pair paternity.[18] Species where males are more brightly coloured than females, like fairy wrens, tend to have higher levels of extra-pair paternity than those where the sexes are similar in appearance, like Hunter's fulmars and other seabirds. One explanation for this pattern is that in the dim and distant past females preferentially sought extra-pair copulations from males that were relatively attractive. As a consequence, selection favoured, and continues to favour, bright plumage in males because it increases their reproductive success. So, it seems that sexual selection mediated by sperm competition has been responsible, in part at least, for the difference in the appearance of male and female birds.

Let us now return to the once cherished notion of female sexual monogamy. Although female sexual monogamy – which, from now on, I'll simply refer to as female monogamy – is the exception rather than the rule, examining the conditions under which females are genuinely monogamous can be enlightening. Monogamy can occur either because a female chooses to remain faithful to one male, or as a consequence of a particular lifestyle.

We can think about fidelity in terms of the advantages it provides to females, and there are some female invertebrates for which remaining faithful is apparently their best strategy. Males and females of some worm species pair for life – literally, fusing their genitalia together before sexual maturity and remaining bonded for ever after. One advantage of such an arrangement is that they never need worry about finding a sexual partner again.

A particularly apposite example of monogamy involves a devastating dipteran pest of cattle, the screwworm fly. The female fly lays her eggs on wounds, large or small, and the larvae feed by consuming the living flesh. Some idea of how awful this must have been for cattle is provided by the accounts of a few unfortunate humans who became infected. A man who had eggs laid in his nose described his initial symptoms as being similar to those of a severe cold. But as the fly larvae ate their way through the tissues of his head his symptoms became increasingly severe and he became more and more miserable. He eventually died when the larvae destroyed his soft palate and entered his eustachian tubes. Screwworm larva cannot be much fun for cattle either and the damage they wreaked in the early years of this century ran to millions of dollars in the southern United States. An effective means of control was much needed.

In a fortuitous but brilliant insight, Edward F. Knipling, an entomologist with the US Department of Agriculture, decided to use the screwworm to try out what was then, in the 1930s, a new method of biological control: the sterile insect technique. Knipling worked out from first principles that if it was possible to rear enough sterile males and release them into the population at the start of the breeding season, females would fail to reproduce and in just a few generations numbers would decline dramatically. Knipling assumed, along with everyone else since Darwin, that female screwworm flies were monogamous. All his calculations of the number of sterile males to release were based on this critical assumption, and by a remarkable bit of luck it turned out to be true: the screwworm fly is one of the few insects where females copulate only once in their life.[19]

Knipling's efforts were a spectacular success, and the screwworm has been reduced to such an extent that its impact is now negligible. Not surprisingly, the sterile insect technique became a bandwagon for those trying to control other insect pests, but all subsequent efforts were dismal failures – precisely because the females were not monogamous. For the females of these other species it made little difference if they copulated with a sterile male: if he couldn't produce viable zygotes one of her other partners would. On the other hand, the research that took place to figure out why the sterile insect technique did not work revealed that most female insects were sexually promiscuous.

In those few instances where it occurs, the implications of true female sexual monogamy are far-reaching – especially for males. For insect species where females copulate just once and are sexually receptive before or soon after they emerge from their pupal cocoon, males are under intense selection to locate virgins. Being first is what counts or, if not quite first, being big enough to displace possible competitors. Among the giant saguaro cacti in the Arizona deserts lives a rather dull-looking bee with no common name, known only as *Centris pallida*. This is a solitary nesting species and females lay their eggs in specially constructed underground cells. In the spring the males emerge first and after gorging themselves on pollen they zoom off in search of females. Males do not wait to encounter free-flying females, but instead seek those struggling to emerge from their underground chamber, dig them out and copulate with them. Once a male starts to dig towards a female, his actions invariably attract other males and the result is a battle for possession. The largest male invariably wins, and once he has copulated with the female he releases her and goes off in search of another.[20]

The race for males to be first in species such as this generates such a strong selective pressure that inseminating females as they emerge into adulthood is widespread. In some insects males have taken this a stage further and inseminate females while they are still wrapped in their pupal shrouds. Among certain butterflies, males locate female pupae, penetrate their hard outer casing with their penis, and inseminate them.[21] The ultimate example of 'being first' is provided by a tiny, primitive insect known as a thrips. Males locate females before they have even started to pupate, and then inseminate them. Females that are immature in every other respect therefore carry sperm with them throughout their final phases of development, ready to lay fertile eggs as soon as they emerge into adulthood. Presumably, early insemination carries no net cost for females, otherwise they would have been under strong selection to defer their receptivity.

Among fish, female fidelity is rare, occurring sometimes by circumstance and sometimes by choice. Let's start with circumstance. The biology of several species of deep-sea angler fish probably dictates that females are monogamous. These fish live at depths of around 2000m in perpetual darkness and have the dubious privilege of being one of the few vertebrates in which males are parasitic and much reduced in size. The females are sedentary sit-and-wait predators. Using their long luminous angling lure, which sprouts from their head, they entice scarce and unsuspecting prey to their death. Male angler fish permanently attach themselves to the underside of a female from an early stage in life. Little more than bags of sperm, males extract nutrients by fusing their tissues directly with those of the female and tapping directly into her blood supply. The fusion between male and female is so complete that the author of an early account[22] speculated that 'one may almost be sure their genital glands ripen simultaneously; and it is perhaps not too fanciful to think that the female may possibly be able to control the seminal discharge of the male to ensure that it takes place at the right time for fertilization of her eggs'. The most reasonable explanation for this bizarre marriage is that in the lightless ocean depths it is incredibly hard for male and female to locate each other and so that once they do it is in both their interests to remain together. Several adaptations facilitate their initial meeting: larval males are highly mobile, have a well-developed sense of smell and huge eyes. For their part females are perfumed and possess light-producing organs which signal their availability to males. Once they meet, male and female are fused for life. Females benefit by being certain of a fertilization partner and males benefit in two ways: free lunches for life, and since in most cases each female has but a single parasitic male, a high probability of being the sole father of a female's offspring.[23]

Amateur aquarists commonly keep a little fish known as the Bronze *Corydoras* catfish. They rarely get to see it reproduce, which is a pity since its mode of fertilization, which ensures the female's fidelity, is remarkable. At spawning time in the wild the male and female of these 7cm-long fish circle each other for a few moments before arranging themselves in 'T' position with the female's mouth latched on to the male's genital opening. Moments later the female lays her eggs, catching them deftly in a pouch created by her pelvic fins. Close examination of the eggs reveals that they have been fertilized, but how? Japanese biologists[24] answered this question by releasing a tiny drop of blue dye into the water beside the male's genital opening just at the moment when the fish got into the 'T' position. To their amazement, the dye was sucked into the female's mouth, only to emerge, ten seconds later, from her genital opening and on to her eggs. It turns out that the female swallows the male's sperm, passes them through her gut and deposits them over her newly laid eggs. This utterly bizarre mode of fertilization may be an adaptation to life in fast-flowing streams, which is where these catfish occur in the wild (and why aquarists rarely get them to breed); such a strategy simultaneously ensures that a female gets her eggs fertilized and maximizes a male's likelihood of being the father.

Among the strangest of all fish are the seahorses and pipefish. Small, distinctly unfish-like in appearance and in many parts of the world seriously endangered, their breeding biology is unusual in that the males take sole care of the offspring. At spawning time the female seahorse transfers her unfertilized eggs into the male's brood pouch using a tube-like ovipositor formed by her ventral fins. Only when they are safely inside does the male release sperm directly into the pouch and fertilize the eggs. There is simply no chance that another male could get his sperm into the brood pouch. In some pipefish species, instead of having a brood pouch, the male has an external brood plate on to which the female sticks her eggs. In these cases fertilization is external, but the male can secure his paternity by releasing sperm when no other males are about. The way this happens in one particular species is particularly strange. His genital opening lies below the brood plate, so to fertilize the eggs he ejaculates and at the same time stops swimming so that he falls gently through a cloud of his own sperm. Recent molecular studies of both seahorse and pipefish families have confirmed their complete sexual monogamy.[25]

Birds were once considered to be models of monogamy. It has always been obvious to the most casual observer that most bird species breed in pairs, with a male and a female working together to rear offspring. However, it was assumed that social monogamy also meant sexual monogamy. But, as detailed behavioural and parentage studies of different species began to reveal, the females of most socially monogamous birds are not sexually monogamous at all. Sperm competition and extra-pair paternity are widespread.[26] There are, however, a few instances where female fidelity is the rule. These tend to be those species with life-long pair bonds, where the male and female don't differ much in size or appearance, and where parental care by the male is crucial for producing offspring. Swans epitomize the popular view of monogamy, immortalized in Shakespeare's lines: 'And wheresoe'er we went, like Juno's swans/Still we went coupled and inseparable.'[27] It was once thought that swans paired for life and that if one pair member died the bereaved partner would either spend the rest of its life in celibacy or expire from a broken heart. Bereaved swans do not in fact remain mateless, but nevertheless behavioural observations and paternity studies have shown that, as long as they are paired, female swans are completely faithful.

In the majority of mammals males are polygynous; very few species are socially monogamous. The first species in which both social and sexual monogamy were verified in the wild was the California mouse. It had previously been suspected that this mouse might be truly monogamous; males and females live together in long-term relationships and, unusually for a mammal, males even help to rear the young. David Ribble of the University of California at Berkeley set out to check this in a study he undertook in the beautiful oak parkland of Hastings Natural History Reservation.[28] Mice are nocturnal and notoriously difficult to watch, so Ribble used a cunning indirect method to see who was associated with whom. He dusted female mice with fluorescent powder, knowing that males and females usually shared a nest and slept in close contact. He found that males only ever bore dust of a single female, strongly suggesting sexual monogamy, a fact that he later confirmed by DNA fingerprinting.

Very occasionally sexual monogamy can be convincingly established without resorting to molecular methods. The hamadryas baboon provides a case study – indeed, this is one of the most singular examples of social and sexual monogamy among female mammals. The hamadryas baboon is a typical polygynous primate: troops comprise several breeding males each paired with one to three females. It was thought that females *were* faithful, because they had never been seen to copulate with a male other than their main partner. But, as the studies of supposedly monogamous birds showed us, what you see is not necessarily what you get, and failing to see infidelity doesn't mean it doesn't occur. However, in this case it was true. Female hamadryas baboons remain totally faithful to their male and this was confirmed accidentally in the course of another study.

In many areas of Africa, and especially around tourist camps, baboons can become confident thugs, intimidating humans and stealing food. I once experienced this at close range in Botswana and found the aggressive behaviour of large males distinctly scary. When baboons become a pest the usual solution is for park rangers to shoot them, but more humane attempts to control them have been made. In one experiment French biologists tried to emulate Knipling's screwworm scheme, by sterilizing male baboons – in this case by vasectomizing them. In an isolated group they vasectomized four of the five harem leaders and then followed them over the next four years. During this time the two females paired to the intact male produced six offspring, but none of the six females paired to vasectomized males produced any offspring at all.[29] This result is remarkable because the females must have been well aware that they were not conceiving, yet they still did not copulate with the intact male. This suggests either that natural infertility among males is so infrequent that females have not evolved an appropriate response to it or, more likely, that the aggressive guarding behaviour of females by their males evolved in the context of sperm competition to ensure that female baboons are never tempted to copulate elsewhere.

Although I have described several cases in detail, sexual monogamy is unusual. In the majority of species, right across the animal kingdom, the general pattern is for females to copulate with more than one male or, among external fertilizers, to have their eggs fertilized by more than one male. Paternity studies confirm this for many invertebrate groups and every major vertebrate group: fish, amphibians, reptiles, birds, marsupials and placental mammals. The only major gaps in our knowledge are the echidnas and duck-billed platypus, but it will probably not be long before we know whether females of these egg-laying mammals are sexually monogamous or not.[30] The consequences of polyandry for males are considerable, and in the next section we explore the way female copulation patterns have shaped male sexual behaviours.

Protecting Paternity: Behavioural Adaptations to Sperm Competition

In 1770 Gilbert White wrote to his naturalist friend Thomas Pennant, about birds pairing up in spring.[31] We still talk about birds pairing up, and the usual explanation for this behaviour is that it is the manifestation of the pair bond – the social glue between male and female. This is naïve. The majority of birds certainly rear offspring as pairs, but the proximity between the sexes that leads us to think of them pairing up has less to do with a mutual relationship than with males selfishly protecting their paternity. The best way a male can make sure he is the father of a female's offspring is to prevent her copulating with any other male. And one of the most effective ways he can do this is to remain close to her during the time she is fertile. The behaviour in which a male follows a female in order to protect his paternity is referred to as mate guarding.

Across the entire animal kingdom mate guarding is one of two main male behavioural adaptations to sperm competition. The other is frequent copulation, since by inseminating more sperm than any other male a female's partner maximizes his chances of fathering her offspring. Both paternity guards will be most effective when females are fertile – when their eggs are potentially fertilizable. The term 'potentially' is important here because the females of many species store sperm for days or even weeks prior to using it to fertilize their eggs. The timing and intensity of these two paternity guards depend, therefore, on when females are fertile. Which paternity guards males employ hinges on their relative costs and benefits. If following a female around seriously interferes with a male's ability to forage, then frequent copulation without guarding might be the better option. Alternatively, if copulation renders individuals vulnerable to predators, then mate guarding and infrequent insemination might be the optimal combination.

Behavioural forms of paternity protection are most easily studied in those animals which can be readily observed: mainly birds and insects. One of the earliest written accounts of mate guarding was made by Edmund Selous in the early 1900s.[32] He described how, as he watched a pair of blackbirds build their nest, the male followed the female, 'hopping where she hops, prying where she pries ... for the cock is as busy in escorting and observing the hen as she is in collecting material for the nest'. Somewhat overshadowed by his more famous brother, a big-game hunter, Edmund Selous was remarkable both in terms of his skill as a field ornithologist and in having the explicit objective of testing Darwin's ideas about sexual selection. Although he was years ahead of his time in both respects, Selous never made the connection between males following females and protecting their paternity.

It was a further seventy-five years before the first quantitative study of mate guarding in a bird was undertaken. For a period of seven or eight days, starting a few days before the female lays her first egg, male sand martins appear to be obsessed by remaining close to their partners. Every time she leaves the nest he follows. The period over which this close proximity is maintained coincides closely with when females are assumed to be fertile – and is therefore consistent with the idea that a male's proximity reduces his chance of being cuckolded.[33] Similar patterns were subsequently reported for many other birds but, as some researchers pointed out, this closeness might not necessarily constitute mate guarding. A male, for example, may remain near to his partner because he's hoping she will allow him to copulate or he may remain close to be able to warn her about predators. The best way to determine whether the male's proximity constitutes mate guarding is to conduct an experiment and establish whether extra-pair copulation attempts are more frequent when the guarding male is absent.

This is precisely what David Westneat did in his study of red-winged blackbirds.[34] Male red-wings are polygynous and keen to keep their fertile females in view or to remain close to them. As in other bird species, proximity is maintained by the male rather than by the female, albeit not as assiduously as by some other birds, presumably because male red-wings often have to guard several fertile females simultaneously. When Westneat captured males and kept them off their territory for one hour the rate of intrusions and extra-pair courtships by males from neighbouring territories increased a hundredfold. Fertile females were no more likely to leave their territory in the absence of their partner; so it appeared that extra-pair behaviour was initiated by male red-wings alone. Subsequent DNA fingerprinting of the offspring in the nests of females whose partners had been temporarily removed revealed considerably more extra-pair paternity than in other nests. This experiment beautifully confirmed that a male's proximity to a female during her fertile period really did protect his paternity. Finally, Westneat asked why, if guarding was so important for males, did they ever leave their territory when they had fertile females? Westneat suspected that males were sometimes forced to relax their mate guarding simply to find enough to eat. This proved to be true: after Westneat provided additional food in some territories, males spent much more time at home *and* fathered a greater proportion of their brood.

The amount of time a male spends guarding a female depends on how long she is fertile for. In birds a female's eggs are fertilized individually at intervals of about twenty-four hours, and guarding spans several days, starting just before the first egg is laid and ending once the last one is fertilized. In other animals a female's eggs are fertilized more rapidly and guarding is correspondingly shorter. In the yellow dungfly male guarding is confined to a brief period immediately after insemination during which time the eggs are fertilized. Dungflies typically copulate for about forty minutes. After the male disengages his genitalia the female immediately starts to lay her eggs into the surface of the dung. The male continues to ride on the female's back, however, for the twenty minutes or so it takes her to deposit all her eggs. He does this in an attempt to prevent other males from copulating with her. If successful, he can be fairly certain the female will have used his sperm to fertilize her eggs – even if she has copulated previously with other males – because of the last male sperm precedence rule (whose mechanism is described in chapter 6).[35] Of course, this rule also means that it is in the interests of other males to supplant guarding males and inseminate their female. And this is exactly what they attempt to do.

In species where the *first* male has the best chance of fertilizing a female's eggs, guarding occurs before copulation. Pre-copulatory guarding is the rule in the so-called freshwater shrimp *Gammarus*. Males use their specially modified forelegs to grasp a female and carry her beneath them, for several days, until she sheds her exoskeleton and is ready to copulate. Since there is no sperm storage and eggs are fertilized immediately, sperm competition is virtually impossible. In this situation mate guarding serves to secure partners rather than fertilizations.[36] To make sure that they get a sexual partner, males acquire females on average about nine days before they are ready to copulate. For an animal with such a tiny brain male-guarding behaviour is surprisingly sophisticated. When the sex ratio is female-biased, male *Gammarus* spend less time guarding, but in the opposite situation, when females are in short supply, males guard for much longer, grabbing females much earlier in their moult cycle. A guarding male hangs on to a female, waiting for her to cast off her old exoskeleton, the only time when her eggs can be fertilized. As soon as she moults the male turns her over and, placing his paired penises inside her brood pouch, inseminates her. Within ten minutes he releases the female and has nothing further to do with her.

Pre-copulatory mate guarding also occurs in many mammals because, just as in *Gammarus*, fertilization usually takes place soon after insemination, but not so soon that sperm competition is precluded. Traditionally, the mate guarding of mammals has been referred to as 'consort behaviour', and depending on the duration of the female's fertile period – oestrus – consortships may last either a few hours or several days. Female ground squirrels are usually in oestrus for just one afternoon each year. Moreover, because they are induced ovulators – that is, copulation triggers ovulation – and ovulation is soon followed by fertilization, males are desperate to be first to find an oestrus female. Once they have found one, they pursue her relentlessly, encouraged by their clear priority. As soon as the female is ready, she allows the male to copulate with her, after which he loses interest in her. Among primates consortships can last several days, partly because oestrus lasts much longer than in ground squirrels, but also because females are spontaneous ovulators and so males are usually much less certain about exactly when ovulation will occur.[37]

Whenever females are polyandrous a male's best chance of fertilizing some of her eggs will depend, ultimately, on the relative abundance of *his* sperm at the site and time of fertilization. Mate guarding is one way to regulate a female's copulation behaviour, but frequent copulation with the same female is the other way a male can increase his chances of fertilization. Much of the enormous variation in copulation frequency in different animal species is closely tied to the intensity of sperm competition: where the risks of cuckoldry are high, a male's best chance of fertilizing a particular female's eggs might be to copulate repeatedly with her.

The frequent copulator award goes to an insect. The swamps and streams of the southern United States provide a home for a variety of formidable creatures, including a number of giant water bugs. At over 10cm long, these creatures are well named. They feed on fish and frogs, injecting an enzyme-rich saliva into their prey and then slurping up the digested juices.

Anyone who carelessly steps on a giant water bug gets the same treatment – which can be extremely painful – and for this reason they are known as 'toe biters'. Unlike most insects, parental care is well developed in giant water bugs. In the back-brooding water bugs females stick their eggs, as they lay them, on to the male's back, and he then carries them around, aerating them and, eventually, helping the young bugs to hatch. The care the male provides is essential since unattended eggs always die. In an evolutionary sense, it is worth a male expending time and energy in caring for offspring only if they are his. And to ensure that the baby bugs on his back really are his the male copulates very, very frequently with the female. Just as in the dungflies we looked at earlier, a water bug's eggs are fertilized immediately before they are laid. The male inseminates the female; she lays one or two eggs and sticks them on to his back. Taking no risks regarding his paternity, the male then inseminates the female again. She lays one or two more eggs, and sticks them on to his back. This cycle of insemination, egg-laying and attachment of eggs goes on and on until the female's entire clutch is on his back. In one case a pair copulated over one hundred times in 36 hours as they transferred 144 eggs. Bob Smith, who conducted this remarkable study,[38] verified that the male's relentless copulation did indeed guarantee his paternity. He performed an experiment in which females were first inseminated by one male and later by another. Paternity markers showed that in every case the second male fertilized over 99 per cent of all offspring. The precise way that the second male achieves this is not known, but it seems likely that by repeatedly inseminating the female he ensures that it is mainly his sperm in the vicinity of the eggs that are just about to be fertilized.

In other animals, such as birds, where males are likely to be cuckolded it seems plausible that frequent copulation will enhance a male's fertilization success. The frequency of copulation in different bird species varies dramatically, from just one to over five hundred times per clutch. High copulation frequencies are common in two categories of birds: those with a polyandrous mating system and those in which mate guarding by following is not practical. Among the former females typically have sexual and social relationships with several males – each one trying to outdo the others in an attempt to win fertilizations – hence the reiterated copulations. The other category of frequent copulators are birds of prey and colonial species. In addition to the absence of mate guarding, these species have one thing in common: substantial parental investment by the male. Mate guarding is impractical because food supplies are often remote and one partner must remain at the nest site to protect it from rivals. And, in contrast to most other birds, care provided by a male raptor or colonial seabird is essential if young are to be raised successfully. Precisely because the male invests so much in rearing offspring he is keen to guarantee his paternity and, in the absence of mate guarding, frequent copulation appears to be the next best thing.[39]

In conclusion, paternity studies have made it clear that sperm competition is widespread. Where there's sperm competition, there are usually paternity guards, and mate guarding and frequent copulation are the most widespread behavioural adaptations to sperm competition. These behaviours have evolved because whenever females copulate with more than one male, it pays males to protect their paternity. These are not the only adaptations to sperm competition, as we shall see, and none of them is entirely infallible. If they were, there would be no sperm competition. The other important point is that we are not dealing with a static system: sperm competition is a dynamic evolutionary force. As soon as an adaptation arises, such as male guarding, it instantly generates a selective force for males to devise a way to circumvent this. In some instances these counter-adaptations are behavioural, but others are physiological and anatomical. To understand how a single sneaky extra-pair copulation could out-compete several inseminations by a female's regular partner, we need to consider several additional factors. These include the female herself (chapter 7) and, as the next two chapters will reveal, the design and operation of the reproductive system of each sex.

3 Genitalia

> *... the Testicles [ovaries] of Women are very useful; for when they are defective, Generation work is quite spoiled ... they contain several eggs ... one of which being impregnated by the most spirituous part of the Man's Seed in the Act of Coition, descends through the oviducts into the womb where it is cherished till it becomes a living child.*
>
> ANON., *The Complete Masterpiece: Displaying the Secrets of Nature in the Generation of Man* (1741)

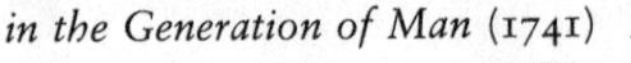

The structure and function of reproductive systems have been sources of endless fascination for anatomists and physiologists. At first sight, understanding the reproductive tract appeared to be little more than an exercise in plumbing: if one could establish the connections between the various components the problem would be largely resolved. By the sixteenth century this basic information was available and it all seemed relatively straightforward: the male ducts transported semen from the testes, down the vas deferens and out through the penis. In the female, the oviduct transported semen one way and eggs and babies the other. However, as research continued, it became increasingly clear that male and female reproductive tracts are much more complex than these apparently simple processes require. In addition, the reproductive anatomy of different species can differ dramatically – a fact exploited by early comparative anatomists to elucidate evolutionary relationships, but one for which there was no obvious explanation.

Part of the problem of understanding why particular reproductive structures and physiological processes have evolved resides in the mistaken assumption that reproduction is a co-operative affair between males and females. The discovery, in the early 1970s, of both male–male competition, in the form of sperm competition, and of sexual conflict between males and females as they each attempted to exploit the other, completely changed our perspectives and how we interpret reproductive features.[1] The objective of this chapter is to demonstrate how the conflicting evolutionary processes of both sperm competition and sperm choice have been instrumental in generating many of the extraordinary features of female and male reproductive systems. Using mainly vertebrates as examples, we shall begin with the female reproductive tract and look at its major roles – the acceptance, storage and utilization of sperm – and how each role has been modified in the face of reproductive conflict. We then do the same for the male system, starting at the testes, working our way down to the sperm stores, digressing slightly to explore the array of accessory glands, and finish by considering the penis.

Egg Makers

Among internally fertilizing species the female usually transports eggs from the ovary to a specific point in her oviduct where fertilization can take place. To facilitate this, the oviduct must also transport sperm, either directly or indirectly, from the site of insemination to the ova. The oviduct does more than act as a conduit for eggs and sperm, however. In species that lay external eggs, like most insects, most reptiles and all birds, the oviduct finishes off the production of an egg which then continues its development outside the female's body. In most mammals, the fertilized egg continues and completes its development in a special part of the oviduct.

The domestic fowl holds a singular place in the history of reproduction since it was the first animal in which the processes of egg formation, fertilization and laying were examined in detail.[2] The hen's single ovary lies like a bunch of different-sized, yellow grapes on the inside of her body wall in the middle of her back. Each 'grape' is an ovum and at full size about the size of a ping-pong ball. The ovum is what most of us would think of as the yolk of an egg. The faint white dot on the ovum (which we can see when we crack open a raw egg) is the germinal disc within which is the female pronucleus – her tightly packed chromosomes. The oviduct is a convoluted tube running from a funnel-shaped region (the infundibulum) next to the ovary down to the outside world. At intervals of twenty-four hours ova are released one at a time from the ovary into the gaping mouth of the infundibulum which is where fertilization takes place (figure 2). Snake-like, the oviduct now starts to pulsate and contract, squeezing the ovum on its way. As it does so the white of the egg (the albumen) is added, and around this a papery membrane provides the foundation for the shell. The final packaging, the formation of the shell, occurs in the uterus region of the oviduct and takes eighteen hours to complete, after which the egg is finished and ready to be laid. Some twenty-four hours after its escape from the ovary the egg is squeezed out through the vagina and into the nest.[3]

The passage of the hen's ovum down the oviduct is only half the story. The transport of sperm up the oviduct provides the other half. In the wild or among free-range chickens, insemination occurs when a cockerel grasps the hen rather roughly by the comb on top of her head and forcibly copulates with her. In this brief and insensitive mounting he transfers about 100 million sperm into her vagina, but as the female picks herself up and walks away she squirts a drop of fluid from her cloaca.[4] With the aid of a microscope you would be able to see that, remarkably, this drop of fluid contains around 80 million sperm. From the male's perspective this seems like an appalling waste, but it is something he has little or no control over. But worse is to come. The sperm remaining inside the female's vagina find themselves in an extremely hostile environment. They swim vigorously to get out of it, seeking the safety of the storage tubules located at the utero-vaginal junction. There are 14,000 sperm storage tubules; sausage-shaped and one-third of a millimetre long. At the entrance to each one, like an orgy of beckoning fingers, are rows and rows of tiny cilia, enticing the sperm to swim deep inside. Once inside the sperm lie motionless, but over the next three weeks they are gradually released

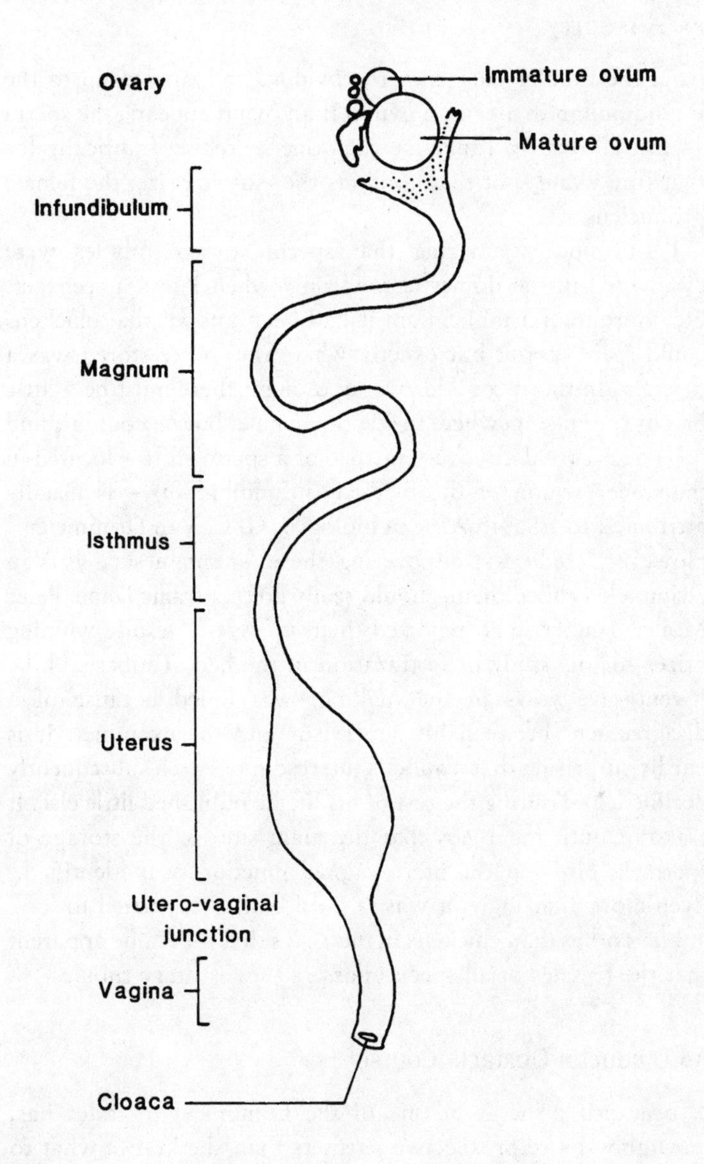

Figure 2

The reproductive tract of a female bird, comprising the ovary and oviduct. The oviduct is divided into a number of regions; sperm are stored in tubules located at the utero-vaginal junction.

from the tubules back into the oviduct and carried up to the infundibulum to await an ovum. If an ovum appears, the sperm swarm all over it. I imagine each one searching frantically for that tiny white spot that contains the elusive prize: the female pronucleus.

I still find it amazing that sperm storage tubules were discovered in the domestic fowl only when I was a teenager. For more than a millennium it had been known that chickens could store sperm, but exactly where they were stored was a mystery. In the 1600s Aldrovandi thought there must be a little bag of sperm somewhere inside the female, but he couldn't find it.[5] The eventual discovery in 1946 of a sperm store – located in the upper region of the oviduct (infundibulum) – is usually attributed to a South African biologist, G. C. Van Drimmelen.[6] However, credit for discovering these 'sperm nests', as Van Drimmelen called them, should really go to a tragic Dane, Peter Møller Tauber, who reported them in 1875.[7] Despite winning prizes for his study of fertilization in the hen, Tauber's Ph.D. (twenty-five years in the making) was failed because of a disagreement between his supervisor and the examiner. It is hardly surprising that Tauber's interest in research subsequently declined, and during the rest of his life he published little else. It was not until the 1960s that the main site for the storage of sperm in birds, at the utero-vaginal junction, was identified.[8] Even more amazingly, it wasn't until biologists started to look at birds other than chickens in the 1980s that it became apparent that the females of all species possess these storage tubules.[9]

An Oviductal Obstacle Course

A beautiful princess in one of the Grimms' fairy tales has, inevitably, more prospective partners than she knows what to do with.[10] To help her decide which to marry she sets them a succession of all but impossible tasks – eventually accepting the one who manages to complete them all. In much the same way the female reproductive tract receives many more sperm than are required; indeed females need only one, or at most a few, sperm to fertilize their eggs. And to make the selection process easier, females have, like the princess, devised a series of challenges. The result is a huge wastage of sperm – an issue that has worried (mainly male) biologists for years and one that lies at the core of sexual conflict. The simple answer to the question of why reproduction is such a wasteful process is that, because of sperm competition, selection has favoured males that transfer lots of sperm since this gives them best chance of fertilization – but there's rather more to it than that, as we'll see in chapter 4. From a female's perspective, the male's seminal exuberance is little more than a nuisance since it requires her to have one or more mechanisms to reduce sperm numbers to a manageable level. Females may have been predisposed to do this anyway because once males started to deposit their semen internally rather than externally, females would have been under strong selection to prevent micro-organisms getting inside their bodies and gaining access to their nutrient-rich eggs. The female's immune system has evolved to identify and destroy foreign bodies and nothing could be more foreign than the sperm cells from another individual. The selection on females to get the right balance between pathogens and sperm would have been immense.[11] An overly vigorous filter that blocked both pathogens and sperm would result in sterility and genes for doing this would go no further. At the other extreme, allowing easy access to both sperm and pathogens would have been dangerous and ultimately another evolutionary dead end. The most effective system for regulating both sperm numbers and pathogens would be a succession of physical and chemical barriers enabling females to reassess the situation at each one.

The female's initial response to insemination in the majority of mammals and birds is to eject most of the ejaculate. Thereafter, the sorting can be more leisurely and more careful. The main filter for birds is the utero-vaginal junction (figure 2). For mammals there are two screening points. In those mammals where semen is deposited in the vagina, the cervix serves this function. For other species, like pigs, which deposit their ejaculate directly into the uterus beyond the cervix, a region known as the utero-tubal junction is the main filter. As well as serving to reduce the numbers of sperm getting to subsequent parts of the female reproductive tract, these constrictions also regulate the flow of sperm. They act as reservoirs from which sperm are released at a slow and steady rate so that there is never more than a few sperm in the vicinity of the unfertilized ovum at any one time. Just as in birds, any sperm that fail to meet an egg pass on through into the body cavity where they are destroyed.

The female tract does more than regulate the numbers and flow of sperm; it also identifies defective individuals, allowing only live and morphologically normal sperm to proceed. This is easily demonstrated in birds. If you take the ejaculate of a cockerel and examine a drop under the microscope, some 10 or 20 per cent of the sperm will show some kind of structural defect: a broken tail, no acrosome or a deformed head. If you now inseminate a hen with the rest of the ejaculate you can later look at the contents of her sperm storage tubules. Under a microscope the sperm can be seen lying inside the tubules: they are all immaculate. No defective sperm have been allowed in. They have been left behind in the vagina to be destroyed or ejected the next time the female defecates. This doesn't mean to say that all the sperm in the tubules are absolutely perfect; some may contain invisible genetic flaws, but at least all the obviously defective sperm are out of the game. A similar process occurs in mammals: immotile or abnormal sperm fall at the first hurdle.[12]

The barriers in the female reproductive tract that sort and control the flow of sperm have led some biologists in the past to suggest that something more sophisticated was going on. One suggestion was that only a tiny proportion of the sperm a male ejaculates is actually capable of fertilization and that the role of the female tract is to identify these special few. The results of some clever experiments performed by Jack Cohen, then at the University of Birmingham, England, were consistent with this idea. After allowing female rabbits to copulate he later recovered those few sperm which got as far as the top of the oviduct. Cohen then re-inseminated these sperm, but mixed with the sperm of a rabbit of another genotype so he could establish which one fertilized the ova. What Cohen's experiment showed was that the sperm from the original insemination fertilized the eggs. Sperm that were good once were good a second time round.[13] This provided support for the idea of a few super-sperm and also a female system capable of identifying and promoting super-sperm.

However, subsequent research suggested exactly the opposite and, rather than there being a few super-sperm, virtually all sperm are capable of fertilizing and producing a viable embryo. The most extreme evidence for this comes from infertile men – whose semen contains no sperm whatsoever. With the help of a sophisticated bit of technology, known as intracytoplasmic sperm injection (ICSI), even infertile men can become fathers. ICSI isn't as much fun as normal sex since it involves inserting a needle directly into one of the testicles to remove a tiny plug of sperm-producing tissue. From this, one or more sperm are extracted and injected directly into an ovum taken from the man's partner. Although the technique is controversial, few of the three thousand babies born so far as a result of ICSI show any more defects than you would expect if you fertilized eggs with the sperm from naturally infertile men. These results indicate that even when sperm have to be dragged screaming and kicking out of the testis itself, every one is a winner, or at least a potential winner.[14]

Of course, what distinguishes Cohen's study from the ICSI result is that in the latter the sperm do not have to negotiate the female tract. What makes those few successful sperm in Cohen's studies is their ability actually to get to the site of fertilization. Cohen's findings caused both controversy and confusion, not least because other researchers struggled to repeat his results.[15] However, in the following decades other findings provided a possible explanation for Cohen's extraordinary results. The most striking involves a plant-like, single-celled organism known as *Chlamydomonas*, normally found in freshwater ponds, but which has long been used in laboratory research. Remarkably, under the right conditions *Chlamydomonas* was able to penetrate hamster eggs. Cohen recognized that if something as unexpected as a protozoan could get through the female tract, then rather than the female tract actively selecting a few fertilizing sperm, perhaps the sperm that get to the site of fertilization are simply those the female cannot 'see'. It has been known for a long time in female mammals that the immune system is closely involved with the regulation of oviductal sperm numbers. Immediately sperm are introduced into the oviduct the vast majority are coated with immunoglobulins, thus allowing the female to 'see' and 'recognize' them as 'foreign' and then destroy them. In Cohen's view, a tiny proportion of sperm in an ejaculate is immunologically neutral, escapes being coated and hence is invisible to the female: these are the ones that get to the site of fertilization. However, Cohen's explanation continues to be controversial and still leaves much to be explained. Why, for example, are 'immunologically invisible sperm' so difficult for males to manufacture?

It is clear from what we have discussed so far that, as well as serving as a conduit for sperm, the female reproductive tract also regulates sperm numbers. One reason it does this is to minimize the risk of polyspermy: the simultaneous penetration of the ovum by several sperm. If this occurs, the resultant embryo fails to develop and dies. There is therefore very strong selection on females to regulate the number of sperm that interact with their ova. As we have seen, in most instances males inseminate many more sperm than are required so females are forced to reduce their numbers. But the female reproductive tract is also sufficiently sophisticated to be able to deal with the opposite situation. If a male inseminates relatively few sperm, amazingly the female can recognize this and increase the number retained. In the following section we look at how females store and utilize sperm.

The Oviduct as a Sperm Storage Site

Some female insects can make a single insemination last years, although how they keep sperm in a viable state for so long is still a mystery. The type of structure that females use to store sperm varies dramatically across species. Some are like the miser who stores small amounts of money in lots of different places: under the bed, inside the wardrobe, in the kitchen cupboard. Birds use this strategy and have large numbers of tiny sperm-storage structures. Lizards, their nearest relatives, are the same, and I guess dinosaurs may have been similar, too. Other animals use the bank-vault strategy and keep their sperm in a single place, in a bag-like storage structure, just as Aldrovandi (erroneously) imagined in chickens.[16] This is the situation in insects whose sperm store – the spermatheca – delivers a few sperm to each egg as they pass down the oviduct. Whether or not a species has one or many sperm-storage sites probably depends on the number of sperm needed to fertilize each egg. Birds and lizards, with their large ova and tiny fertilization target, require huge numbers of sperm and are best served by lots of tiny storage tubules. Insects, on the other hand, require only a few sperm since these can be delivered rather precisely to the micropile, the tiny hole through which the sperm enter the egg.

Mammals are the losers when it comes to sperm storage. Their sperm are extremely short-lived in the female tract, surviving for less than forty-eight hours in most cases. If this were not true, the rhythm method to avoid conception in humans would be even less successful. The maximum sperm storage ever reliably reported in humans is about five days.[17] In contrast to other animals with longer durations of sperm storage, most mammals have no specific storage structures, although, as we have seen, in different species the cervix and the utero-tubal junction serve to trap and slowly release sperm over one or two days.

Bats provide a striking exception. Female bats can keep sperm in a viable condition for several months and they seem to be able to do this without any specific sperm-storage structures. The sperm simply fill up the female's reproductive tract at copulation and remain there in a quiescent state until they are required for fertilization. The explanation for prolonged sperm storage among bats is that it is a time-saving device.[18] Bats in temperate regions of the world hibernate. If you spend half the year asleep you have to squeeze a breeding cycle into the other six months. Most bats copulate in the autumn before they hibernate. Females then store sperm over winter and fertilize their ova in early spring, so that the young are born early in the summer. Interestingly, male bats occasionally exploit the over-winter period of female sperm storage, rousing themselves from their midwinter sleep to copulate with torpid females. The results of a study of the noctule bat, which is unusual in producing twins (usually non-identical), revealed that in about 15 per cent of cases the twins had different fathers. On the basis of the number of different males which had contributed to the stored sperm in the female's uterus (derived from molecular analyses of stored sperm) no more than 40 per cent of offspring should have been full sibs. The fact that 85 per cent of twins were full sibs suggests either that the sperm of one male was more competitive than that of the others or that the prolonged storage of sperm by female bats provides them with an opportunity to use one in preference to others.

How is it possible that sperm, single cells with virtually no energy reserves, are able to remain in a viable state for so long? Prolonged sperm storage in the female tract may be an attribute of the sperm themselves; they may be nourished by the female in some way, or it may be a combination of both male and female effects. In some animals, like bats, it is thought that the stored sperm are attached to the epithelial lining of the female tract and in this way obtain sustenance from the female. In birds, the fluid-filled storage tubules may provide nourishment for sperm. But, in general, we know virtually nothing about the mechanism of prolonged sperm storage.

Why Store Sperm?

The main consequence of sperm storage is that it uncouples the three main reproductive events of copulation, fertilization and birth.[19] Imagine a hypothetical species in which fertilization always occurred twenty-four hours after copulation and birth always occurred nine months after fertilization. The three reproductive events are inextricably linked and birth always occurs nine months after copulation. If there is an optimal time for young to be born then copulation must occur about nine months prior to this. If there is an optimal time to find a partner and to copulate that *isn't* nine months prior to the optimal birth season, there's a problem. One evolutionary solution to this is to break the link between insemination and fertilization by storing sperm. The same effect can also be achieved by either delaying implantation of the fertilized egg or regulating the rate at which the embryo develops. Under delayed implantation, a very early-stage embryo is held in a state of suspended animation until it is embedded in the wall of the oviduct. With delayed development implantation occurs but the development of the embryo is arrested at some stage. Because in the majority of mammals sperm storage is limited to just a few days, most species use delayed implantation to 'uncouple' reproductive events. This strategy allows females to delay the interval between copulation and the start of embryo development by several weeks and, in the extreme case of the weasel, by ten months. One proposed explanation for why delayed implantation is the preferred strategy in mammals is that their high body temperature is not conducive to sperm storage. This in turn could explain why bats, whose body temperature falls considerably during hibernation, are able to store sperm for so long. On the other hand, the body temperature of birds is generally higher than that of mammals and they store sperm for several weeks, so there must be more to it than this.

Another advantageous spin-off from storing sperm is that it provides females with an opportunity to change their minds. In species in which fertilization occurs almost immediately after insemination a female irrevocably commits her eggs to that male. A female who stores sperm prior to fertilization can modify her choice of father for her offspring. She can copulate with one male and secure a supply of sperm, but if a better male comes along, she can copulate with him and use his sperm to fertilize her egg.[20] Of course, males are unlikely to have been indifferent to this and, as we shall see in later chapters, have been under continuous strong selection to avoid being messed about in this way.

Storing sperm will also be advantageous in situations in which ova are fertilized sequentially over long periods of time, particularly if individual ova can be fertilized only during a short time-window. This is exactly what happens in birds: each ovum is fertilizable only during the twenty minutes immediately after it has been ovulated. For this reason all birds have the ability to store sperm: it ensures that sperm are always available to fertilize ova. Different bird species, however, vary in their ability to store sperm. Pigeons are limited to an average of six days. At the other extreme, turkeys can store sperm for an average of forty-five days and hold the overall avian record with a female who produced a fertile egg 110 days after her last insemination. The idea that female birds store sperm to avoid the necessity of copulating as each ovulation occurs is supported by the fact that the duration of sperm storage is longer in species that lay more eggs. Birds usually lay one egg each day, so larger clutches require longer sperm storage. There is, however, an important exception to this. Albatrosses and other marine birds lay only a single egg yet are able to store sperm for as long as eight weeks. The explanation is straightforward. These birds have to travel vast distances between their offshore feeding grounds and their breeding colony where they copulate. Females spend several weeks at sea feeding and accumulating sufficient nutrients to produce their single, large egg. The male, on the other hand, has to remain at the colony (without feeding) to protect the nest site from other birds. The result is that, rather like shift workers, seabird partners rarely see each other and must make the most of it when they do, and, to ensure that their eggs are fertilized, females store sperm.

Many reptiles have a similar problem. They do not usually form enduring pair bonds, and some species occur at such low densities that the chances of males and females meeting at the right time are remote. Under these circumstances it would pay a female to copulate with almost any male she meets and store his sperm until she is ready to reproduce. However, we do not know whether a low encounter rate is what has favoured sperm storage in reptiles. Most reptiles have prodigious sperm-storing abilities. We know this largely because of herpetologists who, having caught a female in the wild and kept her in solitary confinement, have been surprised when she subsequently produced fertile eggs. Swivel-eyed chameleons and time-worn terrapins routinely store sperm from one year to the next. Several snake species store sperm for two or three years and one female Javan wart snake produced fertile eggs no less than seven years after being taken into captivity. While it is obvious that reptiles have a remarkable sperm-storage capability, undoubtedly associated with their ectothermic (cold-blooded) existence, extreme records must be considered with some care. Had this female wart snake had no chance to copulate during her seven years in captivity? Can the possibility of virgin birth be ruled out? Several lizard species reproduce without sex, but until recently there was no evidence that any snake could do so. A recent study[21] has suggested this possibility – although it remains to be substantiated.

After publishing a popular article on sperm storage in different animal species, which contained mention of the brevity of human sperm storage, I received a letter from a man who thought he might have evidence for a much longer duration of human sperm storage. He told me that he worked on North Sea oil rigs and was away from home for three months at a time. He had come home from his last shift to find that his wife was two months pregnant. He was a deeply religious person, he said, and felt that his partner's pregnancy was either a case of immaculate conception or a record for human sperm storage. I threw away the letter, assuming the writer to be a crank. The next day, however, I got a telephone call from a woman, saying she was the man's wife. She was deeply upset and, between her sobbing, told me how her husband had shown her the letter before he sent it. She was able to phone now, she said, only because her husband was out. She confessed that her pregnancy was neither a record for sperm storage nor anything miraculous. Rather, she had become lonely while her husband was away and the obvious had happened. I was horrified to have become involved in this personal drama and the phone call left me feeling shaky for the rest of the day. Only later did I discover that the whole thing had been a set-up by my graduate students.

Blind Sperm Makers

It is rather disconcerting, but probably not altogether surprising, given he had an illegitimate son, that St Augustine (AD 354–430), whose ideas form the basis of Christianity, thought that God had given men testicles to remind them of the irrevocability of original sin. The view of Paracelsus, writing in the 1500s, wasn't much more encouraging: he felt that God had given men testicles so that they could castrate themselves to become impeccable Christians.[22]

Poor Aristotle was confused about testes despite using a certain logic to ascertain their function. His idea was that everything in Nature was done because it was either *necessary* or *better*. Because they differ in shape from those of other vertebrates, Aristotle failed to recognize the existence of testes in fish and snakes.[23] He knew that these animals produced semen, so he deduced that the testes were not *necessary* for reproduction. By means of a convoluted argument, he concluded that animals with testes were *better* because the testes make the 'movement of the seminal residue more steady' with the result that these species, including humans, showed more restraint in copulation.

Leeuwenhoek was closer to the mark when, in his famous letter to the Royal Society in 1677 on the discovery of sperm, he wrote, 'I have no doubt that the testicles have been made by no other purpose than to furnish the little animals in them, but to keep them until they are ejected.'[24]

The term 'testes' has the same root as the word testament or witness, and derives from the Roman custom of holding the testicles when taking an oath. Rather than use the term 'testicles', which has strong human connotations, I will use the more general term: *testes* (plural) and *testis* (singular) to refer to the male sperm-producers.

The testes of most species are paired structures, spherical or ovoid in shape and lying within, or sometimes, as in the human male, outside the body cavity. They comprise numerous long (seminiferous) tubules which join together at one end. The spermatozoa are formed in special cells lining the walls of the tubules. Just as with the ova, sperm are formed by cell division and contain just half the genetic material of the father. As they are produced the sperm are swept along the tubules into a storage area, the epididymis – of which more later. It has been known for a long time that the testes also did something other than produce sperm. The Greeks were well aware that castrated males eventually came to resemble females in their behaviour. Orchidectomy is the wonderful euphemism used nowadays for the surgical removal of one or both human testes, reflecting the morphological similarity between the subterranean bulbs of orchids and the testes. As well as removing the ability to produce sperm, this operation removes the other source of masculinity, the hormone testosterone. Within the testes the spaces between the seminiferous tubules contain the Leydig cells which produce testosterone and other anabolic steroids.[25] The effects of these hormones are numerous and complex and, to put it crudely, they provide males with their sexual motivation and facilitate their aggression which, ultimately, may together determine their reproductive success.

The testes produce astronomical numbers of sperm. An

average man produces 125 million sperm each day, and 2 million million during a lifetime. Viewed in this way it seems like a limitless supply. Cast your mind back, for a moment, to chapter 1 and Trivers's ideas about how the differences between males and females dictate the nature of sexual selection. A key assumption was that sperm were cheap to produce and that male reproductive opportunities were rarely limited by sperm numbers. For females, on the other hand, their large, nutritious eggs were costly to produce and represented the most important limitation on female reproduction. As a result, Trivers said, males copulate indiscriminately while females are coy. If we compare a single sperm with a single egg of almost any species the sperm is obviously the smaller of the two. But this is not a fair comparison since sperm are not delivered singly but in packages (spermatophores) or ejaculates containing millions. The difference in production costs between an ejaculate and an egg may be less than was once assumed. Behavioural ecologists have been concerned with whether sperm production is energetically costly because theory predicts that what you spend on sperm you cannot spend on something else. Reproductive physiologists, however, think the question is trivial, and Jack Cohen, for example, calculated that on a cell-for-cell basis a single human ejaculate was equivalent to just 5 per cent of daily skin loss.[26] However, this might also miss the point, and while no one disputes the fact that measuring the energetic cost of sperm production is difficult, it might not be trivial.

The assumption that sperm are limitless and cheap to produce is wrong, on two counts. First, as all men at least know, sperm are not limitless since every ejaculation is followed by a refractory period of recovery. The recovery period may be very short in certain species, like sheep and chimpanzees, but even they have their limits. Provide a ram with more than fifty ewes on a single day and some will fail to conceive. Second, sperm may not be cheap to produce. In terms of the total number of cells and extra-cellular products in semen, the ejaculate of most species looks rather trivial. Again there are some obvious exceptions –

the voluminous ejaculates of pigs, and the large spermatophores produced by katydids – but in general an ejaculate isn't much to write home about. Cohen's cell-for-cell analogy may be inappropriate since skin cells may be like iron nails and very cheap to produce, while sperm may be more like miniature sculptures and more costly to produce. The problem is that for the majority of species we have absolutely no idea how much energy is needed for sperm formation (spermatogenesis). One reason why sperm production costs are so difficult to measure is that spermatogenesis often takes place at the same time as other energetically demanding activities, such as competing for females. It is therefore difficult to distinguish the energetic costs of these different activities. There are just two studies of two species whose unique features provide an opportunity to estimate the cost of producing sperm.

The first involves the soil-dwelling nematode worm *Caenorhabditis elegans*. This animal has become in recent years a model organism for a whole range of biologists. Its main claims to fame are its small size (a mere 959 cells) and a short but productive lifespan: 350 offspring in two weeks in laboratory culture. From our point of view its singular attribute is that it occurs both as a hermaphrodite and (more rarely) as a male. Copulation stimulates an increase in sperm production and a comparison of males which have and have not copulated revealed that the lifespan of copulating males was 25 per cent shorter than that of non-copulating males. That this reduction in lifespan was a consequence of spermatogenesis rather than sexual behaviour was demonstrated by comparing the longevity of normal males with that of a mutant with reduced sperm production rates. The mutants survived 65 per cent longer than the normal males.[27]

The second example involves a snake – the adder – whose unique circumstance is that spermatogenesis and sexual activity occur in two separate periods during neither of which the male eats. When male adders emerge from hibernation they are extremely sedentary and lie around basking whenever the sun shines. It is during this period that the testes enlarge and spermatogenesis gets under way. After several weeks they shed their skin, and only then start to search actively for partners. It is generally assumed that courting, copulating and fighting are energetically demanding activities and this is confirmed by the fact that male adders lose weight during this period. What is particularly striking is that the rate of weight loss during this very active phase is no different from that during the sperm production phase, suggesting that the cost of spermatogenesis is not trivial and at least equivalent to the cost of reproductive behaviours.[28]

Testis Size

The Swedish physiologist Gustaf Retzius (1842–1919) was the first to draw attention to the fact that the size of the testes, relative to body size, differs rather dramatically between different primate species. Sixty years later, in the 1970s, Roger Short, now at Melbourne University, noted the same phenomenon. When several great apes at Bristol Zoo were anaesthetized in order to move them to new accommodation, Short took the opportunity to look at their testes. He examined three species and confirmed that those of the chimpanzee were relatively huge compared with those of the much larger gorilla. The testes of the orang-utan lay somewhere in between. As a reproductive biologist, Short knew that the size of the testes was an excellent predictor of the rate of sperm production and, in what was to become a landmark paper, he speculated that the need for sperm was fundamentally different between species. Initially, the difference in relative testis size was interpreted in terms of how many times a male needed to copulate. It was well known that male chimps were highly sexed (hence their scientific name, *Pan*) and copulated with all the females in their group. In contrast, because they lived in smaller groups, the copulation opportunities for male gorillas were much less and so they needed fewer sperm. Later, when Short became aware of Geoff Parker's work, he realized that because *female* chimpanzees copulated with several males, sperm competition was another reason why males needed large sperm supplies.[29] Subsequent studies confirmed that it was the copulation behaviour of females, rather than of males, that best predicted relative testis size. This is beautifully illustrated by birds. The males of many grouse and birds of paradise congregate at special display grounds called leks during the breeding season. Females come to the lek with the sole objective of copulating. However, all females have a similar idea about whom they want to copulate with, and sometimes have to queue up for the favours of a particular male. Under these circumstances an attractive male may have to copulate several times in rapid succession so we might expect selection to have produced large testes. But in fact, compared with most other species, lekking birds tend to have relatively small testes. The explanation is strikingly simple. Because the *females* of lekking species generally copulate only once, sperm competition is all but absent. Males can therefore afford to inseminate a relatively small number of sperm and hence do not require large testes.[30]

Towards the end of one of his first papers on testis size, Roger Short wrote, 'It remains to be seen whether the significance of these simple anatomical clues will be confirmed by examination of a far wider range of species.'[31] He must be gratified by the extent to which his idea has now been supported. Not just in mammals and birds, but in butterflies, fish and reptiles: in all of these, species that experience intense sperm competition have larger testes for their body size than others.

How do we demonstrate the relationship between relative testis size and the intensity of sperm competition? Initially we might assume that the easiest way to think about relative testis size is to express the weight of the testes as a percentage of body weight. This would be fine if, on average, the combined weight of the two testes was a constant proportion of body weight in different species, but they are not. Instead, just as larger animal species have relatively smaller brains than smaller species, they also have relatively smaller testes. There is, however, a simple

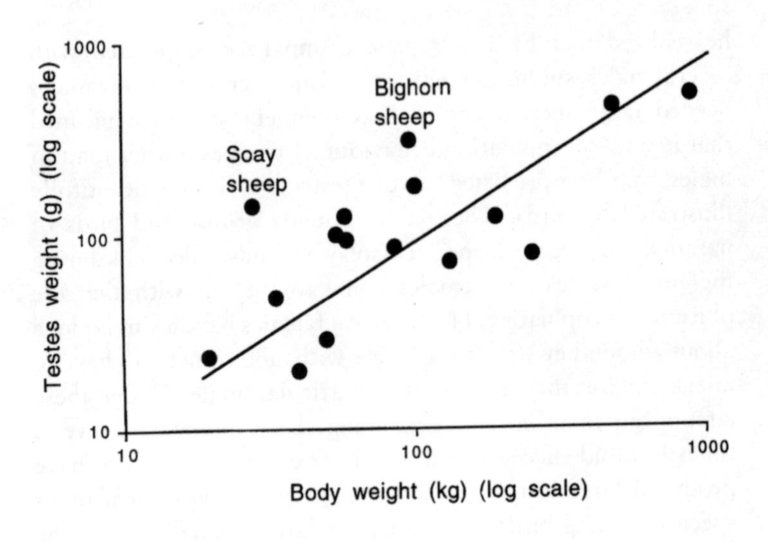

Figure 3

The relationship between body weight and testes weight in 16 species of ungulate. The fitted regression line is shown. The Soay sheep and bighorn sheep have relatively large testes and experience intense sperm competition (from Stevenson (1994)).

way to deal with this. After plotting the relationship between the logarithm of body weight and the logarithm of testes weight, one can simply measure how far above or below the fitted line the points for particular species lie. Points that fall above the line indicate species whose testes are large for the body size and points below the line indicate species with relatively small testes. An example, using ungulates, is shown in figure 3 in which the points for the Soay and bighorn sheep fall well above the line, revealing their relatively large testes, consistent with the intense sperm competition known to occur in these species.

As well as having a measure of relative testis size, we need to be able to estimate or measure the intensity of sperm competition in some way. This has been done for externally fertilizing fishes by considering the number of additional

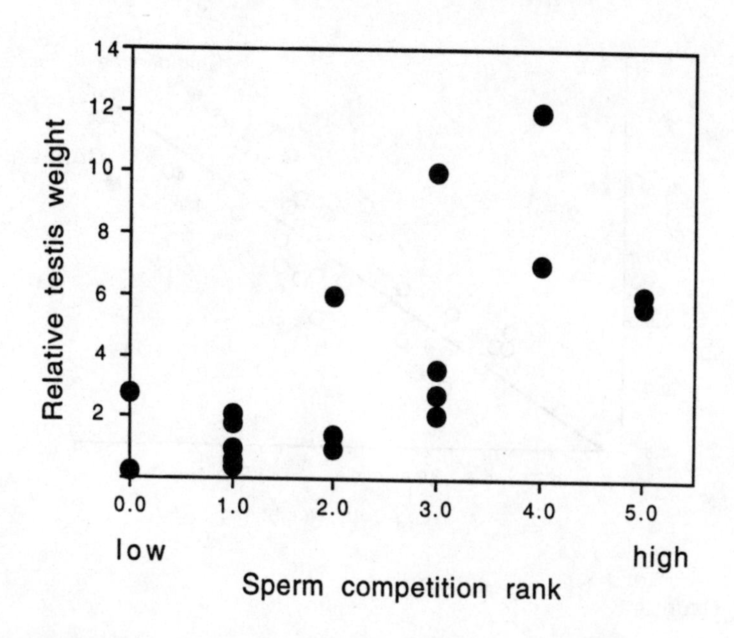

Figure 4

The relationship between the intensity of sperm competition in fish and relative testis size: the more intense the level of sperm competition the greater the testis size (from Stockley *et al.* (1997)).

males that typically associate with a spawning pair (figure 4).[32] There are enough paternity studies of socially monogamous birds to be able to rank the intensity of sperm competition in terms of the level of extra-pair paternity.[33] For mammals we can use the mating system, and whether females are promiscuous, to assess the level of sperm competition.[34] In all cases, intense sperm competition is associated with relatively large testes.

We can use the knowledge that relative testis size across a wide range of animal species is a reliable predictor of the intensity of sperm competition to assess where humans fall in the general scheme of things. The most meaningful comparison is obviously between our own species and other primates and this is exactly what Sandy Harcourt has done.[35] With

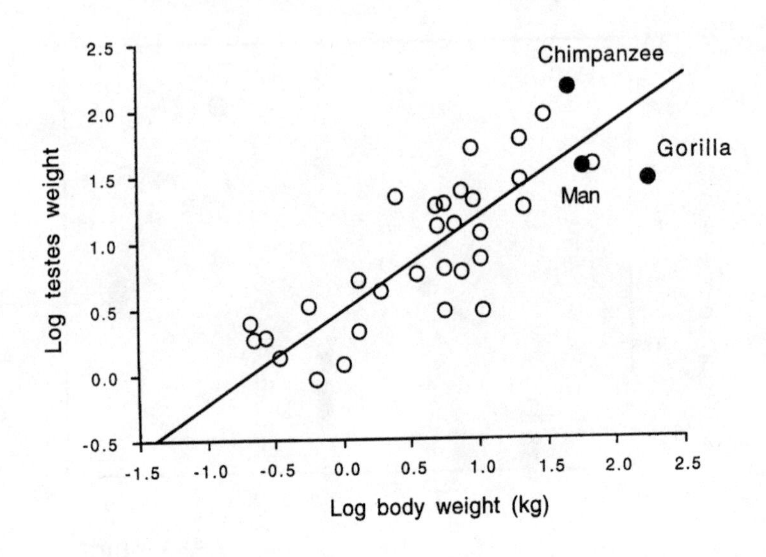

Figure 5

The relationship between body weight and testes weight in primates. The data are expressed as log values. The value for humans falls just below but close to the fitted regression line, indicating that their testes are not especially large or small for their body size (from Harcourt *et al.* (1995)).

information from thirty primate species (figure 5), he plotted the relationship between testis size and body size. At one level the result is clear cut and not too surprising: across all primates bigger species have bigger testes. But if we look at this figure again it is clear that, for any given body size, some primate families fall below the line, indicating that their testes are smaller than we would expect for their body size, while others fall above the line. Among the great apes, gorillas fall below, chimps well above, and humans lie very close to the line, implying that we have an average level of sperm competition for a primate.

Before pursuing this any further it would be sensible to ask whether there is any independent evidence that sperm competition in chimpanzees is more intense than in gorillas, or indeed than in ourselves. We know that female chimpanzees are highly promiscuous, copulating 500–1000 times and with many different males for each pregnancy.[36] And we also know from a remarkable study of parentage in wild chimpanzees that sperm competition is rife. In a study conducted on the Ivory Coast, Pascal Gagneux and colleagues[37] collected DNA from chimps without ever touching them – instead they obtained the necessary DNA from the cheek cells left on chewed fruit, and from hair samples found in nests. Of thirteen infant chimps they tested, no fewer than seven (54 per cent) had been fathered by males from other groups. This finding shattered the long-held assumption that the majority of chimpanzee copulations occurred between group members. They also found that males who consorted with particular females when they were fertile were hopelessly ineffective at protecting their paternity. Only two out of six consorting males fathered the female's infant.

Unfortunately no comparable data yet exist for the gorilla, although studies are in progress. What we do know about gorillas suggests a more modest sexual life with less intense sperm competition. Social groups comprise either a single dominant silverback and a harem of females or, slightly less commonly, two males sharing a harem of females. In the latter case, the dominant male copulates more frequently than the subordinate, but most females copulated with both of them. Copulations involving males from outside the social group have not been observed. Overall, compared with chimpanzees, the copulation rates in female gorillas are relatively infrequent, about two or three times each day or twenty to thirty times for each pregnancy.[38]

What about humans? Despite claims by some researchers that sperm competition is intense in humans (see chapter 1), almost all the evidence, including our relatively modest testis size, points the other way (figure 5). Paternity studies of different communities have provided estimates of extra-pair paternity ranging from 1.4 per cent to 30 per cent. In Britain in the 1980s it was estimated that about 4 per cent of all offspring were the result of extra-marital sex.[39] However, before leaping for the DNA paternity-testing kit or phoning your lawyer, consider this: very few of these studies are published, so their results are not available for scientific scrutiny. The only published studies are those based on the analysis of blood groups and exclusion analyses – that is, who could *not* have been the father – and such estimates are notoriously difficult to interpret. Contrary to expectation, the sophisticated molecular technology now available, which can assign paternity with a high degree of accuracy, has failed to provide good estimates of human extra-pair paternity. Studies of the inheritance of disease, which use molecular methods to double-check genetic affinities between putative relatives, ought to be unbiased sources of such data. But they are not because they rely on volunteers. On being told that the information they provide might reveal true paternity, many would-be volunteers melt away. Those remaining are hardly a random sample and hence provide no basis for an estimate. Perhaps the only source of reliable information, the molecular relatedness studies conducted prior to an organ transplant between putative kin, for ethical reasons have not and will not be published. So what can we conclude? Only that extra-pair paternity undoubtedly exists and that its level in contemporary Western human society appears to be fairly modest compared with that of chimps and many socially monogamous bird species (chapter 2).

If further evidence was required that humans have evolved to deal with rather modest levels of sperm competition, we have only to look at a number of other male anatomical attributes. On all counts, humans compare pretty badly with other mammals (table 1). The rate of human sperm production is lower than that of any other mammal so far investigated. The numbers of sperm stored in the epididymis are also low. Sperm quality is poor compared with other primates: approximately 25 per cent of the sperm in normally fertile men show gross morphological abnormality, whereas in chimpanzees and their diminutive cousins, bonobos, less than 5 per cent of sperm

Table 1
Testis size, sperm reserves, daily sperm production rate, ejaculate volume and sperm numbers in different male mammals.

Measure	Man	Bull	Ram	Boar	Rabbit
Paired testis mass (g)	40	650	500	720	6.4
Number of stored sperm ($x10^9$)	0.42	21	126	104	1.6
Daily sperm production ($x10^6$)[1]	4.4	10	21	23	25
DSP by both testes ($x10^9$)	0.125	7.5	9.5	16.2	0.16
Ejaculate volume (ml)	3.5	6.0	1.25	400	3.2
Sperm per ejaculate ($x10^6$)	350	6900	3000	750,000	640

[1] Per gram of sperm-producing testicular tissue. DSP = Daily sperm production.
(From Mann and Lutwak-Mann (1981) and Amann (1981))

are abnormal.[40] Almost the only aspect in which we excel is in penis size. Compared with most primates the erect human penis is relatively large: in fact, it is on a par with that of the bonobo. As we shall see, there are several explanations for a relatively long penis.

Male Sperm Stores

As well as needing organs for the manufacture of sperm, males require somewhere to keep sperm until they are needed. When a male mammal copulates the sperm are drawn from two stores, the vas deferens and the tail end (cauda) of the epididymis.* The epididymis is also involved in the transport and final maturation of sperm. Just as with the testes, the relative size of the epididymis reflects the intensity of sperm competition. In sheep the tail end of the epididymis is huge and contains a staggering 126,000 million sperm, which accounts for the ram's

* The epididymis is so-called because it lies on the testes (*epi*: 'on') which were once referred to as the *didymi* – from the Greek: 'twins' or 'paired'.

ability to ejaculate thirty to forty times per day – albeit only for the relatively short breeding season. Within primates the comparison between chimpanzees and ourselves is instructive. As we have just seen, sperm competition is intense among chimpanzees whose epididymis contains 4,300 million sperm, and where males can ejaculate once an hour for five hours and use only half of their stored sperm. Men, in contrast, have a more limited capacity and six ejaculations in twenty-four hours is enough to deplete the epididymal sperm stores completely.[41]

Throughout history the testes have been the focus of male reproductive function and the epididymi have been virtually ignored. Presumably this is because each epididymis looks (and feels) like an afterthought stuck on the outside of the testis. The fact that the epididymis is so closely associated with the testis may help to explain another phenomenon – the fact that in adult humans and many other mammals the testes hang externally in the scrotum. The testes start out in the abdomen but during the early years of life they descend downwards, so that by puberty they reside outside the body. Aristotle was well aware that not all animals shared this design feature: the testes of elephants and hedgehogs, as well as whales and seals – he pointed out – were internal, just as they are in birds and most other animals. The question of why some mammals should have external and horribly vulnerable testes has puzzled biologists for a long time.[42] The usual answer is that quality sperm can be produced only if the testes are kept relatively cool. This is true and it is one reason why doctors are so keen to check on testicular descent in boys. Unless the testes descend before the age of about five years the thermal damage to spermatogenesis is irreversible and individuals with undescended testes often develop testicular cancer.

It is not simply a question of internal versus external. Six different degrees of testicular descent have been recognized among mammals[43] (figure 6). Elephants lie at one extreme and their testes are situated deep within the body cavity adjacent to the kidneys. We lie at the other, with our external, scrotal testes.

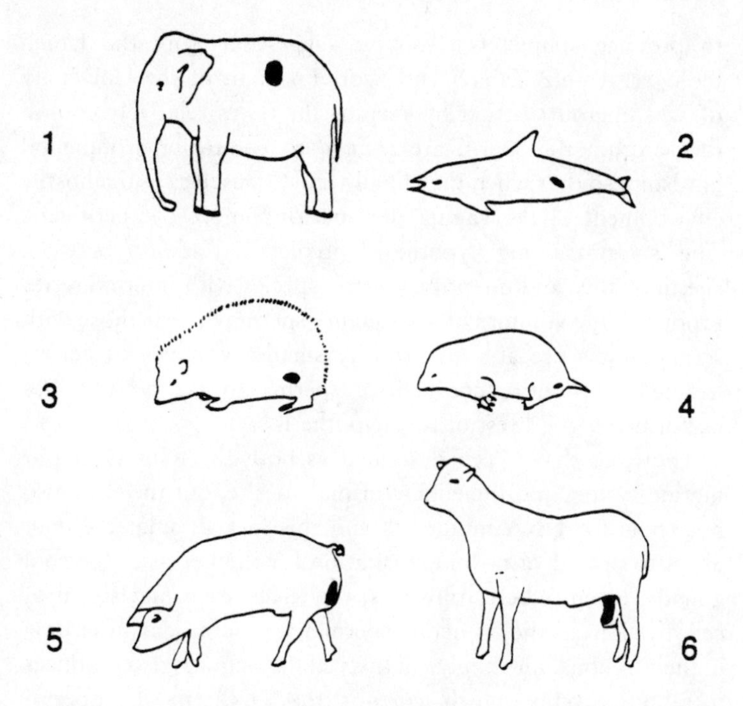

Figure 6

The location of the testes in mammals: six recognized categories of testis location are shown, from the abdominal testes of the elephant to the scrotal testes of the sheep (from Carrick and Setchell (1977)).

The reason for this variation is unclear, but there has been no shortage of ideas. One purely practical suggestion, which I refer to as the 'hard-knocks hypothesis', was made by Michael Chance from the Social Systems Institute at Birmingham. He felt that externalized testes are necessary in animals like us who run and jump because the concussive changes in pressure that occur during vigorous exercise as our insides slop around would forcibly expel the contents of the sperm ducts.[44] An unappealing, inefficient and implausible scenario – to my mind at least. If this was a problem the simplest evolutionary solution would be to produce sphincters. Another suggestion from the Israeli biologist Amotz Zahavi and Scott Freeman of the University of Washington[45] is that by keeping the testes relatively cool in the scrotum the sperm are *trained* to endure environmental hardship, so that when they finally find themselves in the hostile environment of the vagina they perform more competitively. The 'sperm-training hypothesis' predicts a trade-off between sperm quality and quantity, so that species with internal testes produce large volumes of low-quality sperm whereas those with scrotal testes are able to produce smaller volumes of better-'trained' and more competitive sperm. So far no one has performed a good test of this hypothesis.

The testes have been described as both the witness of our mating system and the engine of male desire, but they are also hot spots for DNA mutations and this fact stimulated Roger Short to provide a novel explanation for the necessity for cool gonads. From puberty onwards, the testes are a maelstrom of cell divisions as they generate sperm after sperm, day after day, in their billions. Inevitably, all this cellular activity also produces metabolic toxins which corrupt the DNA in the sperm-producing cells and produce genetic mutations. It is now clear that the rate of mutation in the testes is extraordinarily high and keeping them cool may be one way to minimize genetic damage. Females suffer no such cost – the rate of mutation in ova is a full order of magnitude lower than in sperm – since, as we will see in chapter 5, the DNA in ova becomes quiescent while their owner is still a fetus and remains dormant until the moment of ovulation.

Another view, championed by Mike Bedford, is that the target of selection has been the epididymis rather than the testes.[46] To maintain their fertilizing ability, he argues, stored sperm must keep cool. In humans the scrotal temperature is typically 4° or 5°C lower than inside the abdomen. The importance of storing sperm at a reduced temperature was demonstrated by ingenious experiments on rabbits and rats in which the epididymis was surgically separated from the (scrotal) testes, and deflected back into the body cavity, leaving the testes in the scrotum. After three weeks the effect was dramatic. Despite normal production by the testes, the number of sperm ejaculated was much reduced, presumably because the increased temperature in the epididymis disrupted their maturation, transport and storage. On the basis of this type of evidence Bedford suggests that the epididymis is the prime mover in the evolution of external testes. His argument goes as follows. Sperm obviously *can* be produced by internal testes subject to deep-core body temperatures – as in animals such as elephants and hedgehogs. However, sperm can complete their maturation and be stored effectively *only* if they are kept cool. To this end, the epididymis of the hyrax, elephant and other species with internal testes is located in a cooler location near the body wall (see figure 7). Bedford reasons that in species requiring a large, ready supply of quality sperm, selection has favoured an externally located epididymis. And, during the course of evolution, the testes have been dragged along with it. In turn, the testes of scrotal mammals have evolved to produce sperm at these lower temperatures – hence the problems if they remain undescended.

Although there is still controversy over Bedford's idea,[47] many mammals do show striking adaptations to keep the epididymis cool. These include the remarkable separation of the epididymis and the testes in species such as the hyrax (figure 7), whose testes are abdominal. Even in species with scrotal testes, such as rodents, the epididymis lies at the lower end of the testes, as far away from the body as possible. Moreover, in these species, that part of the scrotum which contains the epididymis is often devoid of hair, so the temperature in this region is reduced even further. As remarkable as these observations might be, the only way Bedford's hypothesis can be properly tested is through a detailed comparison of different species. At present we have insufficient information on the size of epididymal sperm store and the quality of ejaculated sperm and the extent of sperm competition in different mammalian species to conduct a proper comparison.

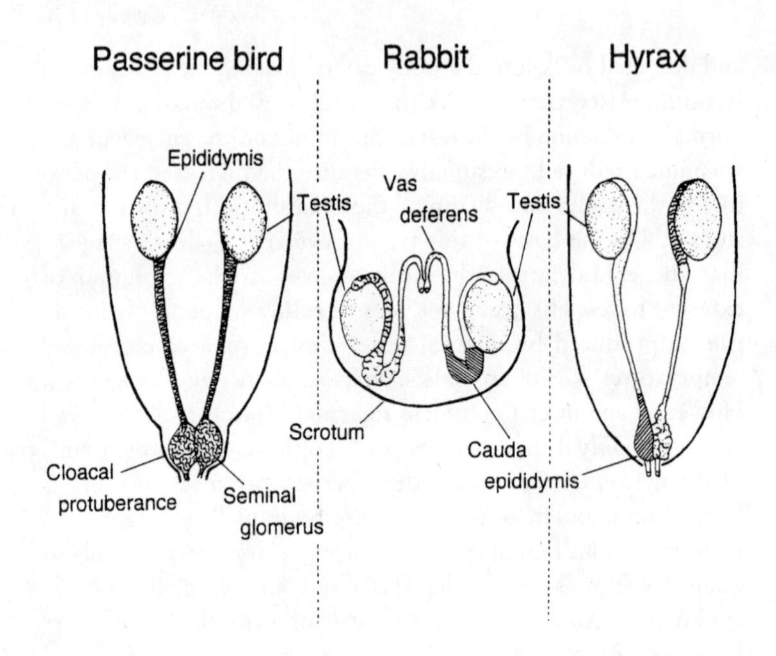

Figure 7

The male reproductive system in a typical passerine bird, such as a house sparrow (left), a typical scrotal mammal, a rabbit (centre) and the hyrax, a mammal with abdominal testes (right). Notice the remarkable similarity between the hyrax and passerine bird (from Glover and Sale (1968)).

If stored sperm must be kept cool, how do birds with their internal testes cope? The answer is immediately obvious if you dissect a male bird during the breeding season. (Spring road-kills provide abundant material for study.) The paired testes can be seen lying on the back of the abdomen wall and running from each of them is the tight and tiny zigzag of the vas deferens. As the vas deferens approaches the cloaca it expands into a region known as the seminal glomerus (plural: glomera). The similarity between a songbird and the hyrax (see figure 7 above) in this respect is remarkable, presumably because the selection pressures have been identical. Each seminal glomerus is effectively the same as the tail end of the epididymis in mammals. Although the term 'glomerus' means 'ball-like', the two wrinkled glomera lying side by side look to me like the two halves of a human brain – something I presume is merely a coincidence.

There's another aspect of keeping bird sperm cool. In the 1930s a biologist[48] studying sperm production in sparrows noticed that, although the testes churn out sperm continuously day and night, the shunting of sperm from the testes to the seminal glomera occurred only at night, when the bird's body temperature was about 4°C cooler than during the day. These observations suggest that sperm are vulnerable to heat damage as they leave the testes and to avoid this are moved rather swiftly at night when body temperature is at its lowest into the relatively cool seminal glomera.

During the breeding season the seminal glomera fill up with sperm and as they do so they push the cloaca outwards to form the cloacal protuberance. For years bird banders have used the presence or absence of this structure to identify the sex of those species in which there are no other clues to gender. Just as with the epididymis in mammals, the seminal glomera provide a cool place in which sperm can mature and are stored. The cloacal protuberance in birds is about 5°C cooler than core body temperature, which the testes experience.[49] The efficacy with which storage occurs in the glomera was nicely demonstrated for me by a male blackbird I picked up dead on the road. I was on a family outing and I placed the bird in the boot of the car. On getting home I put the corpse into the fridge and the next day, almost twenty-four hours after the bird had died, I dissected it and found the sperm to be as vigorous as they would have been in a live bird.

The relative size of the seminal glomera and the cloacal protuberance reflects the intensity of sperm competition in birds even more than the epididymis does in mammals. In 1864 Victor Fatio shot a male small, thrush-like bird in the Alps. When he examined it he was shocked by the enormous size of its cloacal protuberance and wondered why it might need such huge seminal glomera. A century later it was discovered that the Alpine accentor, like its close relative the dunnock, has a polyandrous mating system in which females routinely copulate with several males, resulting in intense sperm competition. The male accentor's seminal glomera are huge because they have to hold 500 million sperm to support the frenzied competition for paternity as males attempt to out-copulate each other. At the other extreme, the similar-sized corn bunting has a harem-based mating system in which females are strictly monogamous. Paternity studies have revealed no sperm competition, and the male corn bunting's seminal glomera contain just 40 million sperm which are used in a modest number of copulations.[50]

Accessories

If you look at a diagram of the reproductive system of a male mammal, such as the rat as in figure 8, you'll see that as well as the usual bits (the paired testes and their epididymis and the penis), there are a number of additional structures. They look like add-ons and they have peculiar names. Just as all roads once led to Rome, the ducts of these glands all lead to the urethra. The secretions of these accessory glands produce the seminal fluid. Prostate, seminal vesicles, coagulating glands, ampullary glands, Cowper's glands, urethral gland, preputial gland – why so many? One reason is that each gland serves a different purpose. Their secretions variously dilute, stimulate, nourish and protect the sperm. In some cases they also produce a vaginal plug and in yet others substances in the seminal fluid may stimulate the female reproductive tract. This still leaves the question of why a single multi-purpose gland did not evolve.

These glands look like afterthoughts because this is exactly what they were. Their existence probably reflects the outcome of successive reproductive conflicts for control over fertilization between male and female, and via sperm competition between male and male. As one sex evolved a better way to wrest control

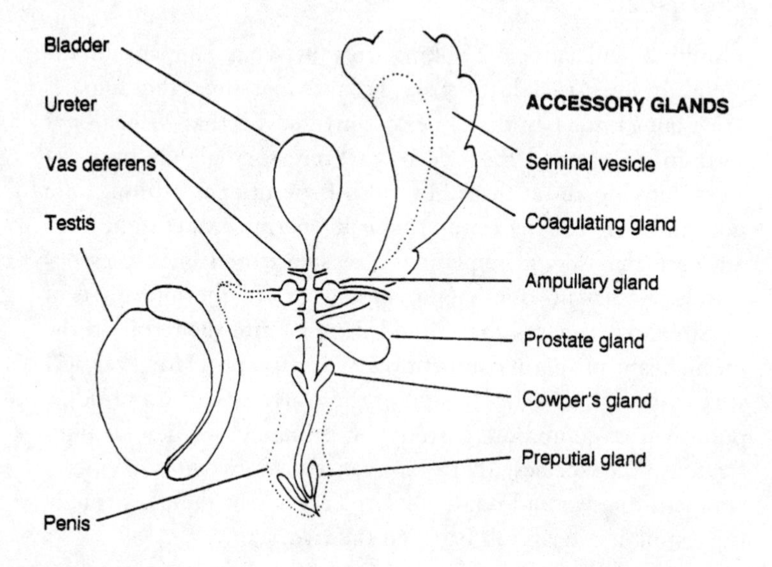

Figure 8

The reproductive system of a male rat showing the six pairs of accessory glands (from Flowerdew (1987)).

from the other, there would be strong selection on the other to produce a counter-adaptation. The plethora of accessory glands might simply be the outcome of this ongoing arms race. It is ironic that despite a large amount of research[51] we still know relatively little about the importance of these different glands in mammals. Some reproductive physiologists even think they have no role at all – for two reasons. First, as we saw earlier (p. 65) when we looked at ICSI, pregnancy can result from the artificial insemination of sperm straight from the testis; that is, without sperm having any contact with accessory gland secretions. Second, animals in which one or more glands have been surgically removed are still capable of insemination and fertilization. But this misses the point. It is analogous to saying we do not understand why males inseminate 10 million sperm when we know from controlled laboratory inseminations that

1 million will suffice. Looking only at what happens in the laboratory misses half the story, because it assumes that females are monogamous. In the same way my guess is that we shall not start to understand the role of the accessory glands until we start thinking about them in terms of sperm competition. I am not implying that their only role is in sperm competition, but I am sure that it will explain the existence of some accessory glands. As yet no one has attempted any of the multitude of possible experiments that would demonstrate their role in the mechanisms of sperm competition (see chapter 6). However, as I was writing this, Alan Dixson, of the University of Cambridge, published a comparative study of primates[52] which demonstrated that species that experienced more intense sperm competition also had relatively large accessory glands, strongly suggesting a causal link between the two.

There are, however, *some* cases where we do know that the products of the accessory glands have a role in sperm competition. They produce the secretions that form copulatory plugs and, as we shall see in chapter 6, they also secrete the potions that knock out rival sperm. These are part of the array of physiological adaptations to sperm competition that supplement behaviours like guarding and frequent copulation discussed in the previous chapter. Copulatory plugs occur in many insects, reptiles, marsupials and in mammals such as certain rodents, bats and primates. In rodents the copulatory plugs are formed from substances that originate from the prostate gland, the seminal vesicles and Cowper's glands. The plug forms in the vagina within a few minutes of copulation taking place. Plugs are often fairly substantial so they must serve an important function. The male guinea pig produces a 3cm-long plug that weighs 1.5g. There have been several ideas about the function of these plugs, including: (i) the prevention of sperm leakage, (ii) the provision of a reservoir for the gradual release of sperm, and (iii) the reduction of the likelihood of another male inseminating the female. The first two hypotheses imply a benefit to both the male and the female. The third hypothesis is concerned with sperm competition and implies that males benefit more than females from the production of plugs. Indeed, in terms of the third hypothesis, a plug may be disadvantageous to a female if it prevents her from being inseminated by more than one male (see chapter 7) and this may explain why female squirrels remove copulatory plugs within minutes of copulation.

Experiments with guinea pigs provided no evidence that plugs serve as a sperm reservoir since unmated females provided with a fresh plug containing live sperm, from another female, all failed to conceive. Other experiments did however suggest that plugs prevented subsequent males from successful insemination. Females who had recently copulated and contained a plug were not inseminated by a second male, but a group of recently copulated females whose plugs had been removed were successfully inseminated and fertilized by a second male.[53] A study of the checker-spot butterfly in the Sonoran desert also showed that the main function of the copulatory plug was to prevent intromission by other males.[54]

The hard, gelatinous copulatory plug in reptiles is formed from substances secreted, not by accessory glands, but by the kidneys. As in the guinea pig, a plug appears to be an effective paternity guard. Many snakes, including the garter snake, hibernate in aggregations of up to 25,000 individuals – investigators sometimes have to stand knee-deep in the snakes they study. Males emerge first in the spring *en masse* and wait for the females. When the females appear the competition between males for copulations is intense and one way males attempt to outdo each other is to plug the cloaca of any female they copulate with.[55]

When I was at school almost every desk had, somewhere on its surface, a crude carving of a heart with an arrow running diagonally through it. This well-known motif, accompanied by a male and a female name, announces who is 'in love' with whom, has its roots in classical mythology: Cupid's arrow. The biological Cupid is the snail, since during its hermaphroditic mutual courtship individuals attempt to skewer each other with a 2–3cm-long calcareous 'love dart' produced in an accessory gland of the female part of the reproductive system. Known for over 250 years, the love dart's function has excited much speculation. One idea was that it represented a nuptial gift. Calcium is important for egg production in snails, so this seems appropriate, but the amount involved is so trivial that this hypothesis seems unlikely. The dart is not propelled through the air but is attached to the owner until it is embedded in its lover. The dart is covered in a whitish mucus, secreted by another accessory gland, and some biologists wondered whether this might be an aphrodisiac – speeding up or synchronizing courtship. But, again, observations failed to confirm this: being skewered had no obvious positive effect on behaviour, and in some cases it even resulted in individuals breaking off their courtship. Instead, the dart-delivered mucus appears to have a more selfish, exploitative role. Injected directly, the mucus manipulates the partner's reproductive physiology, making its reproductive tract more receptive to the owner's sperm.[56] A snail that successfully penetrates its partner with a love dart may therefore increase its own success at the expense of the other – more sexual conflict.

Penis Diversity

The variation in penis shape and form across the animal kingdom is staggering, so much so that certain biologists have made a career out of studying them. The question we are concerned with here is to what extent we can attribute the variation in penis design to sperm competition and sperm choice. We shall start in this section by describing the various types of penis. In the following two sections we shall consider the various uses to which the penis is put, and lastly we shall address the fundamental question of whether sexual selection has helped to shape penis size and structure.

One might be forgiven for assuming that all species with

internal fertilization would possess a penis, but there are several that do not. For many invertebrates and a few vertebrates the spermatophore – a sperm package – is a virtual substitute for a penis. Male cockroaches and scorpions, for example, simply attach their spermatophore to the female's genital opening, from where the sperm leak or are pumped into the female's reproductive system. During an elaborate aquatic courtship ritual male newts place a spermatophore on the floor of the pond and then persuade the female to walk over it and if she does so the spermatophore shoots up into her genital opening. The males of most cephalopods – squid, octopus and cuttlefish – have no penis and also use spermatophores, but one of their many arms is specially modified to serve the same purpose as a penis.[57] This arm is referred to as the hectocotylus and its tip, the tongue-like ligula, is specially designed to grasp spermatophores – taking them from the male's genital opening and placing them inside the female's. As though this wasn't weird enough, the cephalopod spermatophore itself is also remarkable. Once outside the male's body it absorbs sea water by osmotic pressure and then as soon as the spermatophore is inside the female it explodes and releases sperm which the female then stores. The paper nautilus is a rather odd cephalopod with dwarf males whose hectocoytlus doubles as a penis *and* a spermatophore. In this species the male's huge hectocotylus, which is longer than his body, becomes charged with sperm and breaks off inside the female's body.

In the vast majority of bird species a penis is absent, and sperm transfer is accomplished by what ornithologists touchingly refer to as a 'cloacal kiss'. It is generally assumed that most birds lost their penises over evolutionary time – probably as a weight-saving adaptation to flight, for their reptilian ancestors possess one (or, in some cases, two). The few birds that do have a penis include ostriches, kiwis and ducks. In comparison with the mammalian penis, the avian penis is a rather odd structure. The duck's penis, for example, is a strong contender for the design-by-evolution award. During normal business the penis lies retracted inside the cloaca and out of sight, but when a copulation opportunity arises a sophisticated hydrostatic system everts the penis by turning it inside out. Fully operational, the duck's penis is a spiny-textured, spiral-shaped organ extending a full 10cm beyond the cloaca.

One bird has gone a stage further and, uniquely, possesses a false penis. The same size as a starling, the male buffalo weaver possesses a 2cm-long pseudo-phallus directly in front of its cloaca.[58] The false penis is a ductless rod of connective tissue. It is normally invisible, obscured by feathers on the belly, and several ornithologists have studied this bird without even being aware of its presence. Local people in Africa know about the organ and say that the male uses it to carry sticks to the nest. Although there is no evidence at all for this quaint idea, for a time it was the only explanation for the false penis's existence. When I read about the buffalo weaver's unique appendage my first thought was that it might have evolved in response to sperm competition. Its shape suggested that it was intromittent and I wondered whether, like the true penis of some insects, its function might be to remove or manipulate sperm within the female. Together with a research student, Mark Winterbottom, we started a field study of buffalo weavers in Namibia. After three seasons of extremely hot field work and many months in a molecular laboratory back in Britain, we had some idea of what it was all about. First, sperm competition certainly played a central part in the buffalo weaver's life – and in an unexpected way. Instead of breeding as male–female pairs, a buffalo-weaver breeding unit comprised a coalition of two unrelated males and a harem of up to twelve females. DNA fingerprinting revealed that both males in the coalition fathered offspring, but they were also cuckolded by other males. So, sperm competition was intense. Next, we needed to know how the phalloid organ was used during copulation. This proved to be trickier than we anticipated: copulations were difficult to observe and, even on the few occasions when we saw them, it was impossible to see what happened to the phalloid organ. One thing was clear, though: the run-up to copulation and mounting itself were extremely protracted, sometimes taking half an hour. What's more, when the male finally mounted the female he adopted a very odd position, and looked as though he was falling over backwards. Rather than inserting the phalloid organ, the male seemed to do no more than rub it against the female's underside. The other unexpected discovery was that the male underwent an orgasm as he ejaculated – something not recorded in any other bird. Our only explanation for all this is that the phalloid organ has indeed evolved in response to sperm competition. It facilitates male orgasm and ejaculation, but only after an extremely protracted period of foreplay. The advantage to the male may be that the female is more inclined to accept his sperm, or retain a greater portion of it, if she experiences a prolonged copulatory courtship accompanied by manipulation by the male's phalloid organ.

Having dispensed with absent or false penises, let us now confront the real thing. The males of a wide range of animals, both invertebrates and vertebrates, possess a penis and in some cases they may have more than one. Penises come in various shapes and sizes but in most cases the penis is a tube – admirably suited to fulfil its main function of semen transfer. The penis of insects and mammals is a proper tube with a central duct, but in reptiles and the few birds that possess one, the penis is a flat sheet of tissue rolled inwards to create an apology for a tube. Mammals possess one of three basic penis designs. Dogs, pigs, cattle and whales enjoy a penis described as being fibroelastic: it is permanently semi-erect, and its fibrous structure allows its conversion into the pizzles one sees in Spanish tourist shops selling bull-fighting paraphernalia. The second design includes a stiffening bone: the baculum or *os penis*.* A baculum occurs in five orders of mammals: Primates, Rodents, Insectivores, Chiroptera (bats) and Carnivores. I am

* Perhaps surprisingly, the females of some mammals have an equivalent bone, the *os clitoris*.

probably one of only a few people to possess a baculum – one I found lying in glorious isolation at the back of a deep sea cave on Skomer Island where it had once belonged to a grey seal. The final penis design, epitomized by *Homo sapiens*, involves two spongy compartments which fill up with blood during arousal and cause erection. The adaptive significance of the different penis designs has not been explored, but the simplest explanation is that they represent different solutions to similar problems.

Most species have but a single penis. There are some, however, where selection has favoured individuals with more than one. Several marine flatworms have dozens. More modestly, snakes and lizards have the dubious privilege of being endowed with two penises, or hemipenes, as they are known. Each hemipenis is independent of the other, and has its own testis and epididymal sperm store. During copulation only one hemipenis is inserted into the female, and many researchers have noted that males alternate the use of the right and left with successive copulation partners. Detailed study of an *Anolis* lizard from Florida revealed that the alternation of hemipenis use was particularly obvious when the interval between copulations was less than one day. With longer copulation intervals hemipenes were employed at random.[59] This suggested that alternation was linked with the efficacy of sperm transfer, and indeed when experimenters manipulated which hemipenis was used it was clear that by changing over between copulations a male could transfer almost five times as many sperm as if he had reused the original hemipenis. When the interval between copulations was longer than twenty-four hours sperm supplies had been replenished and ejaculate size was similar regardless of which hemipenis was used.

Slugs are the genital giants of the mollusc world, with a penis many times longer than their body. Ironically, their grisly form of copulation means that some individuals end up with no penis at all. Slugs and their shelled cousins, snails, are hermaphrodites, and copulation usually involves the slimy intertwining of two individuals and the simultaneous insertion of penises. In some slugs the penises themselves do the intertwining and sperm exchange occurs outside their bodies. Courtship in the common European garden slug, aptly named *Limax maximus*, involves two animals following each other in a clockwise manner on a vertical surface. On making contact they produce a mucus anchor rope from which they suspend themselves. As the entwined animals dangle in space they each evert and entwine their gigantic penises which far below their owners exchange sperm at their tips. In the slug *Limax redii* the penis is 85cm long – seven times the length of its owner's body. We have no idea whether the slug's giant penis and the unusual pattern of copulation have evolved in response to sperm competition or any other kind of sexual conflict. But we do know that for some species they can be a cause of the worst kind of sexual conflict. Their entwined penises frequently become knotted during copulation and the only way individuals can release themselves is to bite off their own penis at its base! It is not clear whether both individuals have to do this to ensure freedom. If not, there is presumably protracted negotiation about who will sacrifice his masculinity. Thereafter, those individuals that have lost their penis in this way operate only as females.[60]

It is difficult to see how a system as bizarre as this would evolve – who benefits from having his penis severed? However, there is an elusive evolutionary logic. The potential for sexual conflict among hermaphrodites, such as slugs, who cannot fertilize themselves, is very high. Reproduction involves the exchange of sperm and because of this they are very vulnerable to being ripped off, in the sense that one partner may give sperm but the other may cheat and not give any. The recipient benefits by using the sperm either to fertilize its eggs or, more likely, as a source of nutrients. In the latter instance the donor has lost out. Hermaphrodites that cannot fertilize themselves have evolved all sorts of strategies to avoid being exploited in this way. One is the simultaneous exchange of sperm: 'I'll give you some of mine, if you give me some of yours.' This is fine in external fertilizers, but for species with internal fertilization it creates another problem. To achieve simultaneous penis penetration individuals sometimes have to adopt very peculiar positions. Slugs have got round this by exchanging sperm via their extruded penises but, as we have seen, this then creates another problem, which includes the loss of the penis. If an individual bites off its own organ this may be either an error or a last resort (it can still operate as a female). If it bites off its partner's penis this may be a deliberate strategy to minimize the chance that the partner will copulate again, and hence increase the sperm donor's chance of fertilizing the recipient's eggs.

Penis Function

The main function of the penis is the transfer of semen from male to female, but penises also have a number of other roles. In some species the penis serves as a conduit for excretory products, notably urine. On reflection, it seems a less than perfect arrangement to have precious sperm and potentially toxic urine using the same tube. However, this apparently inefficient design has undoubtedly arisen because of the way the reproductive and excretory systems evolved in our distant ancestors. As often happens, our evolutionary past is reflected in our embryological development and if we were able to watch a mammalian foetus develop we would see that the reproductive and excretory systems arise from the same tissue. The end result is a urino-genital compromise – similar to the air–food compromise that results from the oesophagus and trachea using the mouth as a common entrance. Again, this is not the optimal design, merely a consequence of the starting materials available to natural selection. What we now have in the mammalian penis is another evolutionary compromise. It is probably quite a good compromise as these things go since in humans at least we seem to suffer fewer inconveniences from a dual-purpose penis than we do from the air–food arrangement with its ever-present risk of choking. The only shortcoming of the urino-genital compromise is the occasional case of retrograde ejaculation into the bladder. The problem of protecting sperm from the toxic residue of urine was easily solved – by the use of seminal fluids. Although it is not always obvious, semen is ejaculated in rather distinct fractions.[61] In many mammals, fluids from the prostate glands pave the way for the sperm-rich fraction which is thus protected from the toxic effects of urine.

In addition to its role as a way of transferring semen and urine, the penis has evolved a number of secondary roles, some of which are closely linked with sperm competition and sperm choice. These include the removal or repositioning of rival sperm within the female, the removal of copulatory plugs, and stimulation of the female.

Sperm removal is one of the most direct forms of sperm competition and the penis of some species of dragonfly and damselfly[62] is specifically designed for this purpose (see chapter 6). After the discovery in the late 1970s that the damselfly's prickly penis removes competitors' sperm, it was sometimes assumed that any species whose penis bore spikes, spines, barbs, flanges or hooks was into sperm removal. However, this was not the case. Sperm removal appears to be a relatively rare phenomenon. In most instances the purpose of additional penile appendages is to help the male maintain genital contact with the female. This might sound as though spines facilitate copulation for both partners, but again their existence is more likely to reflect a sexual conflict since males can often gain substantially from prolonging copulation. It may allow them to transfer more sperm and other substances to the female which manipulate her behaviour to the male's advantage but to the female's disadvantage (see chapter 4). For this reason there is often an apparent disagreement about how long copulation should last. Females often try to remove a male, by kicking or twisting a long time before he shows any sign of wanting to disengage his genitalia and dismount.

In those species in which males deposit a copulatory plug in the female after copulation, another function of the penis is the removal or destruction of plugs left by other males. One species of squirrel has a knife-like penis which enables it to cut through the plug in previously inseminated females. Similarly, at the base of each of its two hemipenes, the red-sided garter snake has a spine with which it can lever out the plugs from previous males.[63]

Penis Evolution

The huge diversity in penis shape and size in different animal groups has been exploited by taxonomists – those concerned with the classification of animals – to help them discriminate between otherwise indistinguishable species. In many cases, particularly among invertebrates, the differences in genitalic structure provide the only way of distinguishing closely related species. The one question taxonomists never ask is why all this diversity in penis structure should have evolved in the first place. Obviously, as we have seen, some of the diversity is due to the different uses to which penises are put, but this cannot be the whole story. In many animals the penis appears to be used for nothing out of the ordinary, yet the differences in size and structure between otherwise similar species are still considerable. If the primary purpose of the penis is to transfer semen from the male to the female, the enormous variation in penis design seems unnecessarily extravagant.

Two main forces are likely to have shaped the male intromittent organ: competition between males for fertilizations and female choice. Sperm competition has put males under strong pressure to be able to place their semen in the most effective location for fertilization. All else being equal, the longer the penis the closer a male can deposit his semen to a female's eggs. If sperm competition is intense the proximity of semen deposition to the ova may give a male a competitive advantage. All else is rarely equal, however, because it might not always be in the female's interest to have semen deposited deep inside her reproductive tract. Males are interested only in how many eggs they can fertilize. Females, in contrast, are concerned about *who* fertilizes their eggs. This results in sexual conflict, and strong evolutionary pressures on females to retain control over the fertilization process. It therefore seems more than likely that male intromittent organs and female reproductive tracts have co-evolved. Selection on males for a longer penis would be countered by selection on females for a longer reproductive tract. This sexual conflict is one explanation for the sexual swellings that occur in certain female primates when they are fertile. The function of these large, fleshy extensions of the female's external genitalia has long been a mystery. One suggestion is that they increase the length of the female reproductive tract, thereby giving back some control to the female. Another sexual conflict over the location of sperm deposition may explain the variation in penis shape, length and structure in mammals. Pigs provide an extreme example. Male and female wild pigs are promiscuous and so sperm competition seems very likely (although no paternity studies of wild pigs have been published). The pig's penis is exceptionally long and has an unusual corkscrew tip which penetrates and locks into the spiral-shaped cervix, allowing the male to deposit his ejaculate directly into the uterus. The penis is just one of several features of pig reproductive anatomy that have probably evolved in relation to sperm competition. Others include their relatively enormous testes and epididymal sperm store, and an extremely large ejaculate which includes both large numbers of sperm and a huge volume of seminal fluid (see table 1). Overall, across a wide range of other species the match between dimensions of the penis and the female reproductive tract is consistent with the idea that these male and female reproductive structures have co-evolved.

Bill Eberhard of the University of Costa Rica has suggested that of the two possible explanations for penis design – sperm competition and female choice – the latter is the more important.[64] His argument is eloquently laid out in a book I have heard described by some as the most stimulating they have ever read. However, with a title like *Sexual Selection and Animal Genitalia*, it isn't the sort of book you would read on the train. In it Eberhard pits his female-choice hypothesis against the long-established 'lock-and-key hypothesis', which proposed that genitalic congruence between the sexes evolved to prevent individuals reproducing with the wrong species. The lock-and-key hypothesis is unlikely, Eberhard argues, for several reasons, including the fact that there are many instances where males have species-specific genitalia, but females have generalized genitalia and are quite incapable of excluding males of other species. Eberhard's novel view is that elaborate penises have evolved to stimulate the female during copulation. A female then assesses the quality of this internal courtship to decide whether or not to use that particular male's sperm to fertilize her eggs. In other words, the male's performance allows a female to discriminate between him and other males and, having done so, she exerts a cryptic form of choice over whose sperm to use (see also chapter 6). Obviously, no conscious decision is implied here: a particular degree of stimulation simply triggers a female's acceptance of that male's sperm. In this scenario the diversity of penis form has evolved through what can be described as a 'runaway' process of sexual selection, which works in exactly the same way as the evolution of the peacock's tail. Females prefer to be fertilized by males who provide the most stimulation during copulation. The result is the evolution of more elaborate penis structure, which in turn makes females increasingly discriminating, and leads eventually to increasingly elaborate penises, and so on and so on, with penis structure 'running away' until it becomes so elaborate it is more of a hindrance than a help. At this point natural selection steps in and the process grinds to a halt.

A comparative study of the penises of monkeys and apes provides some support for Eberhard's novel idea. Primate penises vary considerably in several different respects: in their dimensions and whether they contain a baculum, the presence or absence of spines, and in the elaboration of the distal penis. Those species in which females routinely copulate with several males during a single reproductive or oestrus cycle tend to have a much more elaborate penis than species in which females were less promiscuous.[65]

Even more convincing evidence, however, comes from a study of insects by Göran Arnqvist of Umeå University in Sweden.[66] As Arnqvist points out, the lock-and-key and the sexual-selection hypotheses make opposite predictions. For females of species that copulate with only a single partner it is crucial that they do so with the correct species. The lock-and-key hypothesis therefore predicts greater genitalic diversity in monogamous species. Eberhard's sexual-selection hypothesis, on the other hand, predicts more genitalic diversity in polyandrous species. Arnqvist compared the complexity of male genitalia and, as a control, other non-genitalic features in monogamous and polyandrous insect families and genera. There was no difference in the diversity of non-genitalic traits between the two mating systems, but almost without exception the male genitalia of polyandrous insects were more complex than their monogamous counterparts, providing compelling evidence that it is sexual selection and not natural selection to avoid inter-specific copulations that has been largely responsible for this diversity.

To conclude, this chapter has looked at the anatomical machinery responsible for two processes: making and moving male and female sex cells – gametes. It is clear that the structure and function of male and female genitalia are much more complicated than these processes require, and that this complexity is a consequence of sexual conflict and competition. In the next chapter we look at how the gametes themselves have been moulded by these evolutionary forces.

4 Sperm, Ejaculates and Ova

What are we to conclude ... if it be not that those Spermatic Worms are the occasion of the Generation of all Animals?

NICHOLAS ANDRY, *An account of the breeding of worms in Human Bodies* (1701)

Eggs and sperm – the stuff of occasional miracles and frequent accidents. Although we now take it for granted that sperm and egg must fuse to produce a new being, the road to this discovery was as long and winding as the oviduct itself. This is hardly unexpected given the microscopic size of sex cells and the temporal separation of insemination, fertilization and birth. It is also not surprising that our understanding of male and female roles in reproduction should have fluctuated through the course of time. In Homer's day females ruled supreme in reproduction and pregnancy was thought to result from microscopic 'animalculae' carried in the air which somehow found themselves inside the female. The male's role was unimportant and the concept of paternity unknown since 'man lacked all sense of responsibility for the survival of the species'.[1] The very term 'mother nature' stems from this period in which goddesses were all important and females dominated reproduction. Easter is the pagan legacy of this; named after the goddess Oestrus and celebrated with the ultimate symbols of female fertility – Easter eggs.

Only when the parallel between planting seeds into the womb of mother earth and the impregnation of a female with semen became apparent did the male's role in reproduction assume a special significance.[2] That insemination was known to be an integral part of reproduction is clear from the Greek story of Pasiphae and her husband Minos. Fed up with his persistent infidelity, Pasiphae put a spell on Minos so that in subsequent affairs he 'poured forth in his semen a swarm of poisonous snakes, scorpions and centipedes, which devoured the woman's entrails'.

In his book *Ornithologia*, published in 1599, the Italian scholar Ulisse Aldrovandi commented on the great lustfulness of the rooster.[3] In contrast to other birds like the eagle and sparrow, who 'copulate less frequently and are content with a single partner, the rooster treads his numerous wives fifty times a day.' Aldrovandi also noted that aggression among cockerels was not associated with the acquisition of food or protection of their offspring, but was motivated entirely by the desire to maintain sole control over their females: 'The rooster fights because he does not wish any of his hens to be touched by anyone and he thus performs the function of a wise father, protecting his honour.' The cause of the cockerels' salacity, Aldrovandi suggested, was their 'especially abundant genital semen: since they cannot endure the irritation it produces they hurry towards sexual satisfaction.' The Italian anatomist Fabricius ab Aquapendente (1537–1619) was the first to identify the ovary of the hen as the source of ova, but was unable to transpose the concept to humans because the ovaries of birds and women are so different in appearance. Fabricius taught and greatly influenced William Harvey (1578–1657), whose main claim to fame was discovering the circulation of the blood. The two men differed on a number of points relating to reproduction. Fabricius thought that hens could store viable sperm for an entire breeding season (several months) following a single insemination, but Harvey accurately showed that thirty days was the maximum duration.[4] Fabricius thought that semen stimulated the generative process without entering the egg; Harvey was convinced that embryonic development was initiated by semen penetrating the egg, but without a microscope he was unable to demonstrate this.

The story of the discovery of spermatozoa – literally 'sperm animals' – by Anton Leeuwenhoek (1632–1723), or more likely by his student Johan Ham, is well known.[5] Leeuwenhoek reported to the Royal Society how, by means of his home-made microscope – comprising a single exquisitely ground lens, which magnified 300 times – he had observed in his own semen millions of vigorously swimming spermatozoa. His letter to the Royal Society is somewhat coy: 'What I investigate is only what, without sinfully defiling myself, remains as a residue after conjugal coitus. And, if your Lordship should consider that these observations may disgust or scandalise the learned, I earnestly beg your Lordship to regard them as private and to publish or destroy them as your Lordship thinks fit.' Luckily, the Royal Society thought it appropriate to publish Leeuwenhoek's findings, in which he suggested that it was the minute microscopic creatures swimming in the semen that entered the egg and resulted in fertilization. This was controversial stuff, and some of his colleagues in the Royal Society thought that all that Leeuwenhoek had seen were parasites. After all, Leeuwenhoek had shown the existence of numerous microscopic animals when he examined the scrapings from his teeth!

It wasn't until a further century had passed that another Italian, Lazzaro Spallanzani, a priest cum scientist, provided unequivocal evidence for Leeuwenhoek's spermatozoa hypothesis for fertility. Given its current unease with matters sexual, it seems rather surprising that the Church should have provided Spallanzani with both moral protection and financial assistance in his efforts to establish the role of semen in reproduction. Spallanzani worked mainly with frogs, whose reproductive behaviour had been lovingly described by the Dutch biologist Jan Swammerdam (1637–80) in his *Book of Nature*.[6] During the breeding season frogs

become so eagerly intent on the business of propagation, that they take no care in a manner of their own safety ... The male frog leaps upon the female, and when seated on her back, he fastens himself to her ... and throws his forelegs round her breast ... He most beautifully joins his toes between one another, in the same manner as people do their fingers at prayer ... and closes them so firmly that I found it impossible *to loosen them with my naked hands ... At last the eggs are discharged in the female's fundament in a long stream, and the male ... immediately fecundifies, fertilizes or impregnates them by an effusion of his semen.* As soon as these eggs have escaped from the female body, between hers and the male's hinder legs, and have been impregnated by the male's semen, the two frogs abandon each other.*

Inspired by some novel but unsuccessful experiments of two colleagues, Spallanzani made pairs of prophylactic oilskin trousers for male frogs to prevent their semen from reaching the female's eggs.[7] The experiment worked: despite the encumbrance of the trousers the males grasped the females, whose eggs were not fertilized, 'for want of being bedewed with semen'. Spallanzani then conducted the other essential part of the experiment. Recovering the drops of semen from inside the trousers, he applied these to a female's eggs which subsequently developed. Moving swiftly from external to internal fertilization and from frogs to dogs, Spallanzani performed the ultimate experiment. He took a female spaniel and before she came on to heat placed her under lock and key inside his apartment, away from male dogs. When she was obviously in oestrus Spallanzani found a male spaniel 'which furnished me, by spontaneous emission, with nineteen grains of seed, which were immediately inseminated' into the female. Sixty-two days later 'the bitch brought forth three very lively puppies' which resembled both the male and female. This was the first ever successful artificial insemination involving internal fertilization. With some justification Spallanzani was delighted by his efforts: 'The success of this experiment gave me a pleasure which I have never experienced in any of my philosophical researches.'

* A recent study revealed that despite their tight amplectic embrace sperm competition is widespread in frogs, with over half of all clumps of spawn having more than one father. Sperm competition must be facilitated by the high degree of breeding synchrony coupled with the dense aggregation of spawning individuals.[8]

Spallanzani's studies demonstrated unequivocally for the first time that semen was essential for fertilization, and in doing so dispelled the centuries-old concept of spontaneous generation. Notwithstanding these clever experiments Spallanzani still thought that 'spermatic worms' played no role in fertilization. The reason for this was the outcome of another ingenious investigation in which he filtered semen in order to establish which component of semen – sperm or seminal fluid – was responsible for fertilization. A mixture of filtered semen and eggs generated fertile eggs and Spallanzani deduced, entirely logically, that it was the seminal fluid rather than the spermatozoa that triggered development. What he had not realized was quite how difficult it was successfully to separate sperm from seminal fluid and it is now obvious to us that some sperm must have remained. On the basis of these experiments Spallanzani believed that the seminal fluid stimulated the foetal heart, which lay pre-formed inside the egg, and triggered development.[9] It was nearly another century before George Newport in 1853 showed, again using frogs, that sperm actually penetrated the egg and were essential if fertilization was to occur.[10]

Spallanzani was an 'ovist', believing each egg to contain a pre-formed embryo. In rather vigorous contrast, the 'spermists' thought that the sperm contained the entire embryo and that copulation and insemination were little more than embryo transfer. In the spermist's scheme the female was regarded merely as a recipient vessel to provide the optimum environment for the embryo's growth. Nicolas Hartsoeker (1665–1725) encapsulated the spermist's view of the male's central role in reproduction in his drawing of a sperm containing an extremely cramped homunculus with a gigantic head. In fact, Hartsoeker never claimed to have seen the little man inside a sperm, merely that if he could this is what it would look like. Nevertheless, the idea of a pre-formed body inside each sperm was an appealing, and not unreasonable, one during the seventeenth and eighteenth centuries. It did, however, worry James Cooke, an English doctor, who wondered in 1762 what happened to all the sperm that did not give rise to a new person.[11] He thought they might not die, but 'live a latent life, in an insensible or dormant state, like Swallows in Winter, lying quite still like a stopped watch when let down, till [they] are received afresh into some other male Body of the proper kind'.

But it was all these wasted sperm, together with their minute size, that finally brought about the demise of the spermist viewpoint. The French physician Pierre-Louis Moreau de Maupertius summed up the spermist's problem in 1744: 'This little worm, swimming in the seminal fluid, contains an infinity of generations, from father to father. And each [pre-formed creature inside the sperm] has his seminal fluid, full of swimming animals so much smaller than himself.' Sperm within sperm within sperm . . . on and on into infinity. Hartsoeker tried to calculate how small the sperm in the original rabbit would have to have been to account for all the rabbits that had ever lived. But it didn't add up. Or, rather, it did, but the answer was so incredible, a figure involving 100,000 zeros, that it seemed ludicrous.

Exactly the same problem faced the ovists.[12] The mother of us all was Eve and her ovaries must, like a Russian doll, have contained an endless series of smaller and smaller homunculi to sustain the human race. Hardly a likely scenario. There were other objections: the ovist view could not, for example, account for the occurrence of hybrids: if the ovum of a horse contained a pre-formed horse, where did mules come from?

The alternative to the pre-formationist view of both the spermists and the ovists was epigenesis – the idea that embryos resulted from the fusion of male and female sex cells, an idea favoured by William Harvey, among others. But even this had its problems. While the observations of early embryologists were consistent with the generation of new structures arising during development, there had to be some sort of pre-formation to account for the resemblance between offspring and their parents. The transfer of this information, it was deduced, must occur at conception.[13] The turning point came in 1875 when

Oscar Hertwig showed, using sea urchins, that the sperm head fused with the female genetic material inside the egg to form the nucleus of the new being.

Sex Cells

Gender is defined by the size of sex cells an individual produces. Males are identified as the producers of small cells or sperm, and females as the producers of large cells or ova. It may seem unlikely, but even this most fundamental difference between the sexes has its roots in sperm competition. There are various answers to the question 'Why are there two sexes?' and they all assume a similar evolutionary starting point: the gametes of our aquatic, single-celled ancestors were all the same size. There was then selection for the extremes, because middle-sized sex cells were less successful than either big ones or small ones. However, on a fixed energy budget individuals could produce either a few large sex cells or many small ones, resulting eventually in females and males, respectively. The difference in the size of male and female sex cells is referred to as anisogamy. Large sex cells were at an advantage because they contained plenty of nutrients, enabling them to survive for long periods. Small sex cells survived less well but were at an advantage since they could be produced in profusion and secured many more fusions. It has also been suggested that small sex cells were at an advantage because they provided no room for parasites. Because of their greater numbers, most fusions would initially have been between sperm, but their small size and lack of resources would have resulted in poor zygote survival. Fusions between two ova would have had very good survival prospects, but this would have occurred rather infrequently because ova would have been grabbed first by the much more abundant sperm. The only evolutionary stable scenario was fusion between small and large sex cells in which sperm effectively parasitize the resources contained in ova. The rampant competition between sperm for fusions ensured relentless selection for large numbers of tiny sperm and hence mergers with larger ova.[14]

Ova and sperm differ in a number of ways other than size. Whereas in the male sperm are continually replenished, females start life with a fixed quota of ova. Females cannot renew their supply of eggs, so what they start out with must last them a lifetime. In fact the situation is even more extreme than this because many developing ova are culled before they get as far as ovulation. A three-month-old human embryo contains about 6 million potential ova. By birth this number has been reduced to half a million, and by puberty it is reduced still further to fifty thousand. The degeneration continues throughout life and a maximum of four hundred ova are produced during a woman's lifetime. This degeneration, or atresia, of ova is part of an ongoing process in the body of weeding out those eggs not capable of forming a viable foetus long before a sperm ever gets near them.[15] In a few cases, however, the selection of appropriate eggs occurs after ovulation. The viscacha, a South American rodent, is extraordinary in this respect: of the thousand or so eggs a female ovulates during a reproductive cycle, only six implant.[16]

The other major characteristic of ova is their immobility, and one consequence of this is that ova differ rather little in form between species. Superficially, this does not appear to be true since eggs differ enormously in size. However, the main difference lies simply in the amount of nutrients they contain. Among reptiles and birds whose eggs develop externally their nutrient reserves (yolk) are massive. An ostrich and a human are approximately the same weight but an ostrich produces the largest ovum known (indeed, it is the largest known cell), some 120mm in diameter. In contrast, a human ovum is just 0.2mm in diameter. The difference, of course, reflects the external and independent development of the avian egg and the internal maternal nurturing of the mammalian embryo.

Sperm are highly specialized cells whose sole purpose is to deliver a haploid set of chromosomes into the cytoplasm of the ovum. The male's role is simple: produce and deliver. One way of looking at it is that the male is simply a sperm's way of making more sperm. Spermatozoa (and the medium in which they are transported, the seminal fluid) have been subject to a multitude of selection pressures. The end result is one of the most remarkable of all cells – designed to penetrate the ovum and to cope with a wide variety of insults along the way.

The Structure of Spermatozoa

The popular image of a spermatozoon, evinced by Gary Larson's cartoons, is of a tadpole-like structure. This structure reflects the main design features: a head, midpiece and tail. The head carries the genetic material: a unique set of DNA packed in a particularly dense form and referred to as the nucleus. In contrast to other cells, the genes in the sperm nucleus are not thought to be expressed by the sperm itself. The nuclear material is not expressed until it has been fused with a female nucleus in the ovum. The sperm head is capped by the acrosome which contains the enzymes necessary for the sperm to digest its way through the outer coverings of the egg. The sperm midpiece contains the mitochondria, the sperm's energy source, which drives the engine, the sperm tail or flagellum. The flagellum is what helps to propel the sperm from A to B and provides the mechanical thrust to facilitate penetration of the ovum. In spite of their common purpose, the sperm of different species vary enormously.

One of the first people to describe and illustrate the wonderful variation in sperm shape and size was the extraordinarily prolific Swede Gustaf Retzius (1842–1919).[17] His three hundred publications spanned microscopy, botany, zoology, anthropology, travel and poetry. His marriage to a wealthy woman, ambitious for her husband's success, allowed Retzius to indulge his various interests and to publish lavish descriptions of his findings. Famous mainly for his work on neuroanatomy and physiology, his contribution to our understanding of sperm

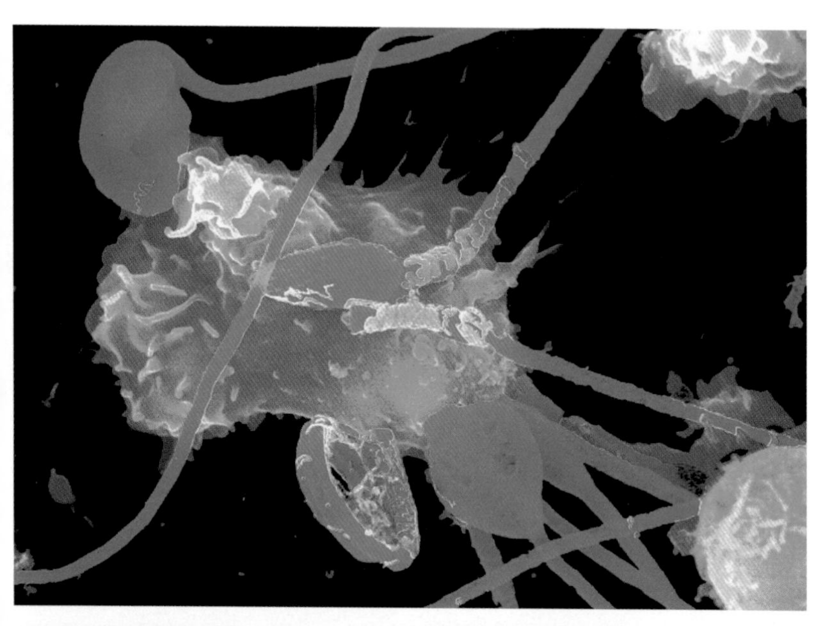

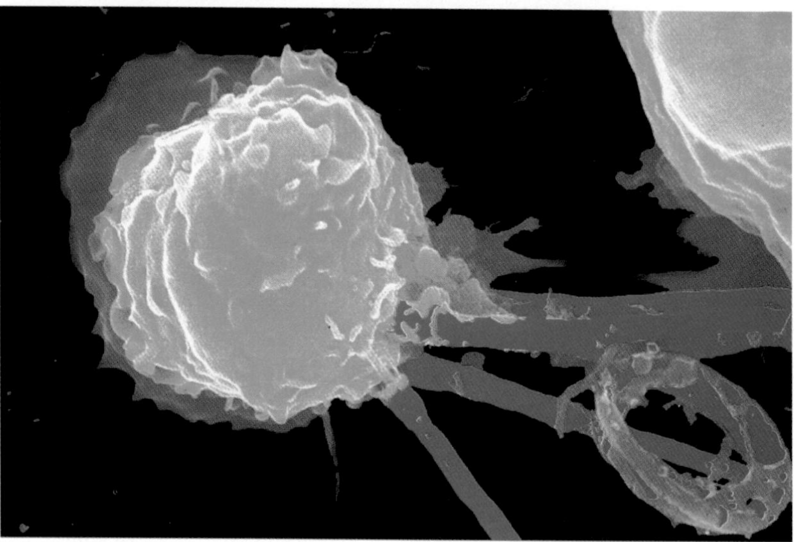

1 The ultimate sexual conflict? Human sperm engulfed and destroyed by a female white blood cell (orange) in the female reproductive tract shortly after insemination. These sperm are failures, the ones that got nowhere near the site of fertilization (chapter 4).

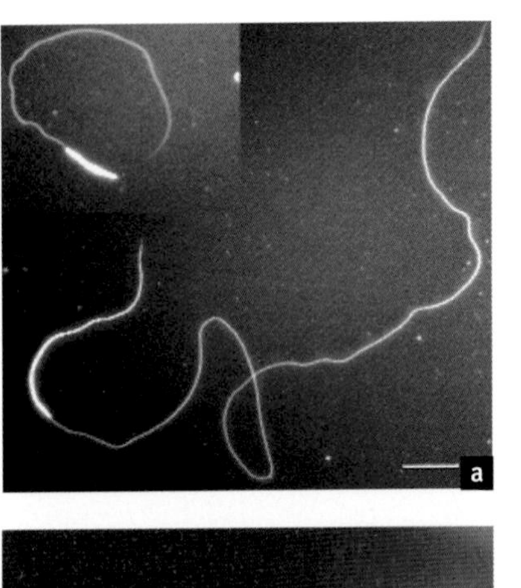

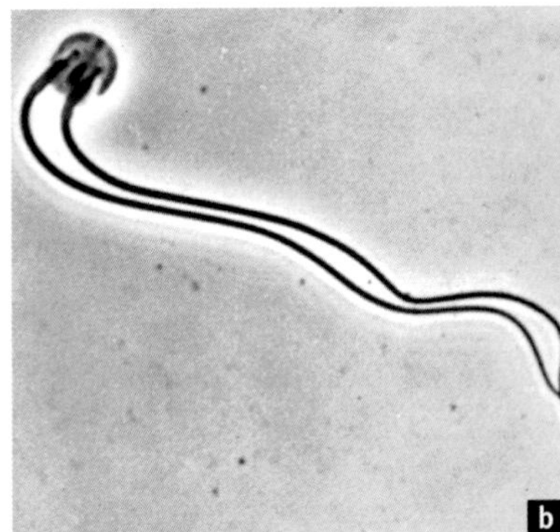

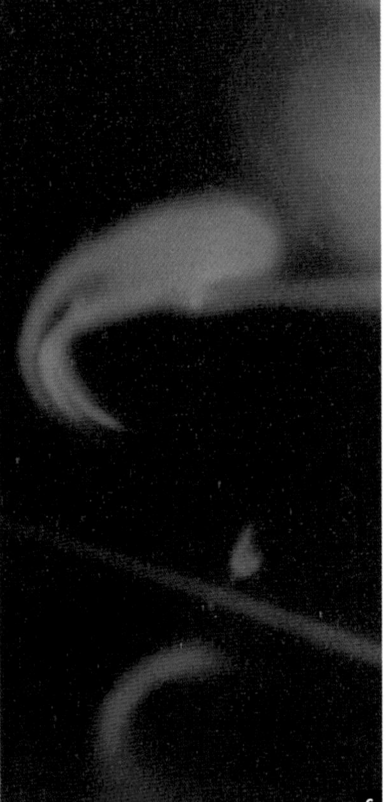

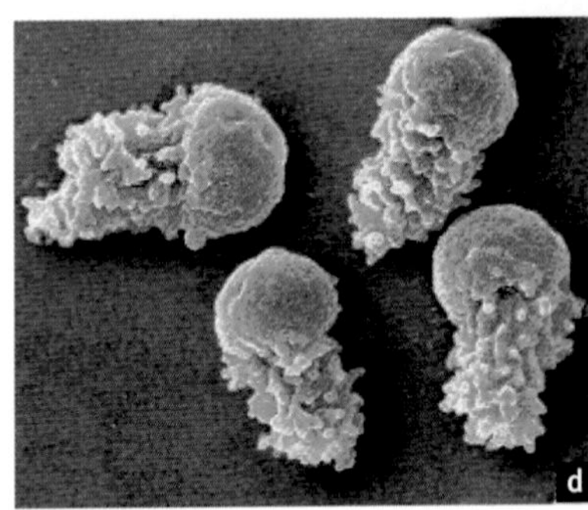

2 A staggering diversity of sperm types (chapter 4). Sperm vary in size and shape both within individuals of the same species and between different species. (a) Sperm from the same fruit fly ejaculate – this species produces both long and short sperm but only long sperm fertilize. (b) Unlikely partners: Opossum sperm operate as pairs for greater efficiency. (c) The most complex of all mammalian sperm is that of an Australian marsupial, the plains rat. This picture shows the remarkable hooked head and fantastic finger-like projections whose function remains a mystery. (d) Improbable sperm: the tailless, amoeboid sperm of a nematode worm — the knobbly processes are 'false legs' which the sperm use to crawl to the site of fertilization.

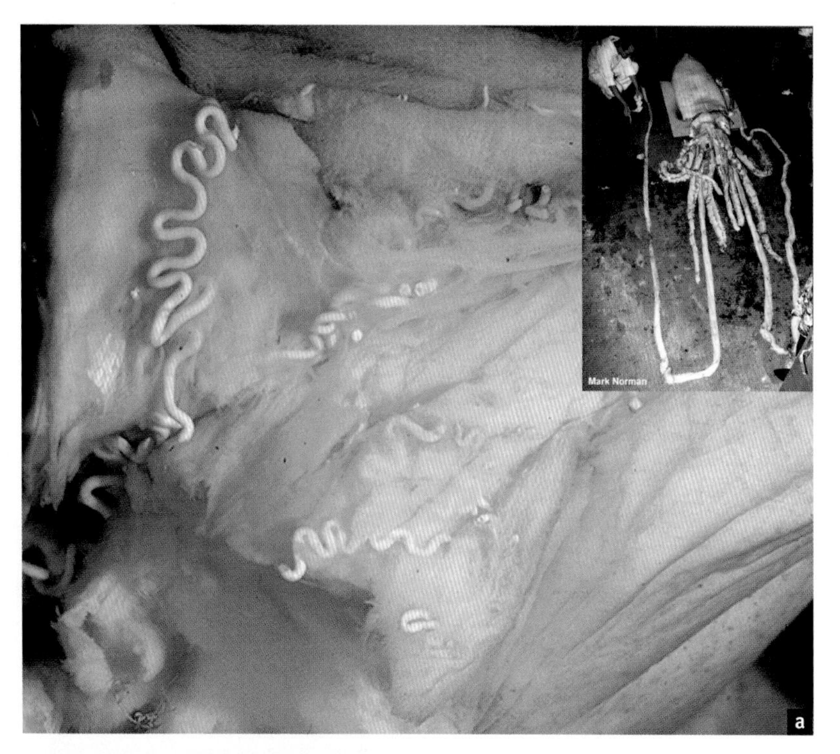

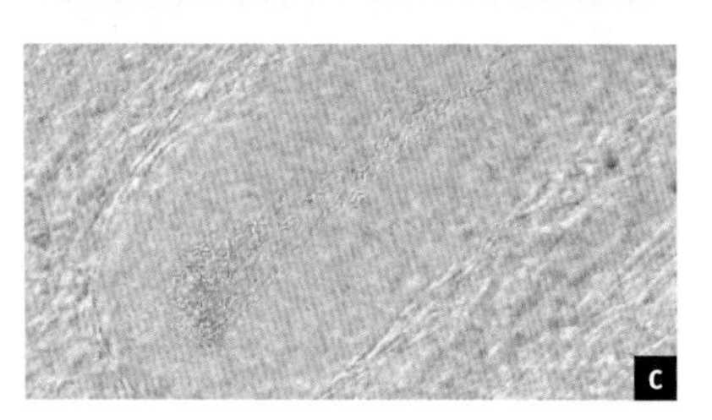

3 Keeping sperm safe — before and after insemination. (a) Sperm packages of the giant squid, injected by the male not into the female's reproductive system but under her skin around the mouth — an eccentric insemination strategy. The inset shows the entire female squid in which these 20cm sperm packages were found (chapter 5). (b) The relatively enormous sperm package of an Australian bushcricket attached to a female immediately after copulation and moments before she starts to eat it (chapter 3). (c) One of seven hundred sperm storage tubules in the oviduct of a female chaffinch — this tubule is packed with sperm.

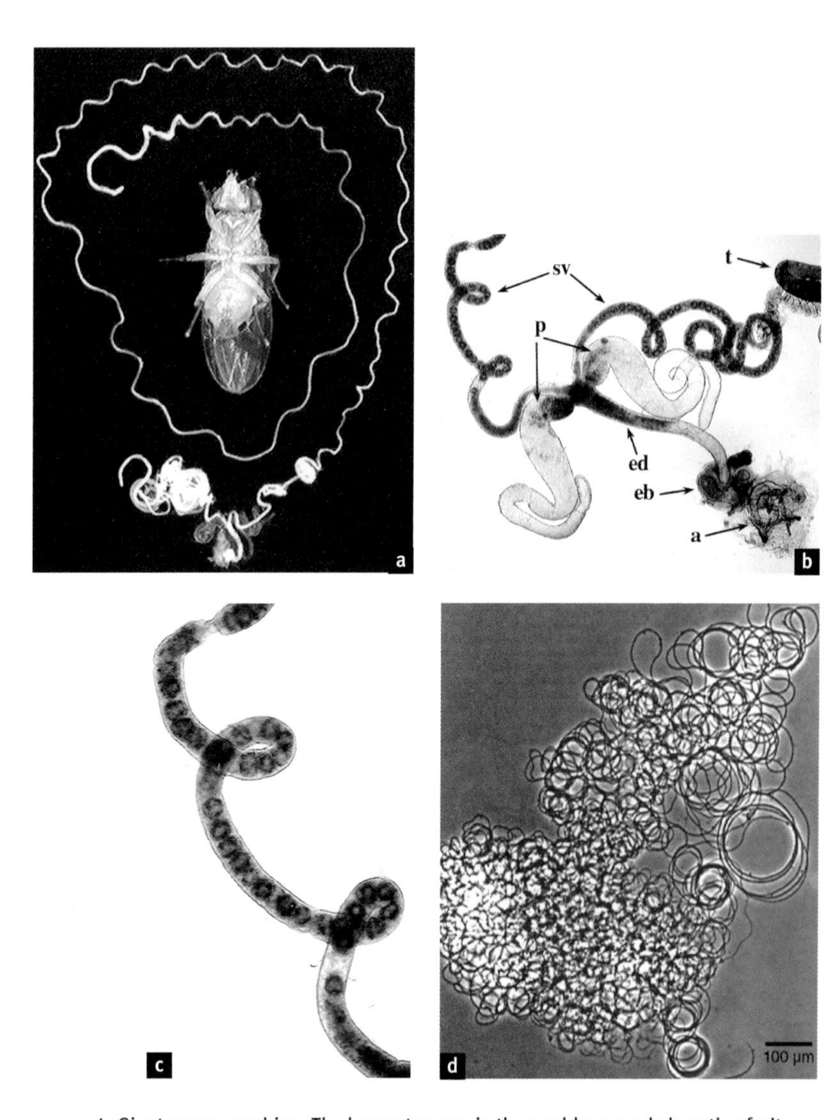

4 Giant sperm machine. The longest sperm in the world are made by a tiny fruit fly. (a) A male fly with one testis (unravelled) from an individual of the same size. (b) The male's reproductive system — showing the relatively enormous paragonial (accessory glands) (p), the male's sperm store — the seminal vesicles (sv), the testis (t), ejaculatory bulb (eb), ejaculatory duct (ed), and aedeagus (penis) (a). (c) A close-up of a sperm store showing the individual sperm rolled up like balls of string waiting to be ejaculated. (d) A single, partially unravelled sperm, 58mm long — many times the length of its owner (chapter 4).

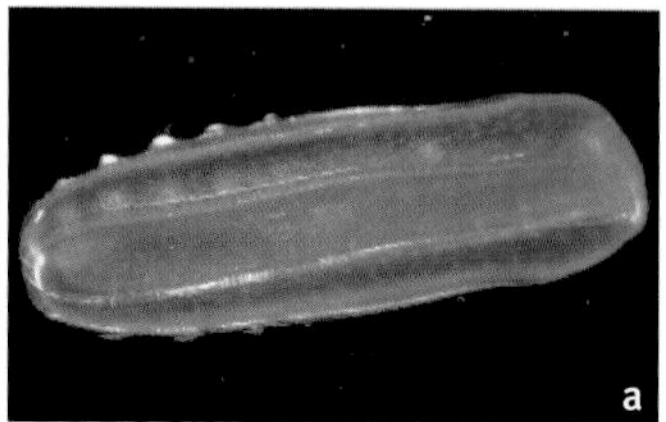

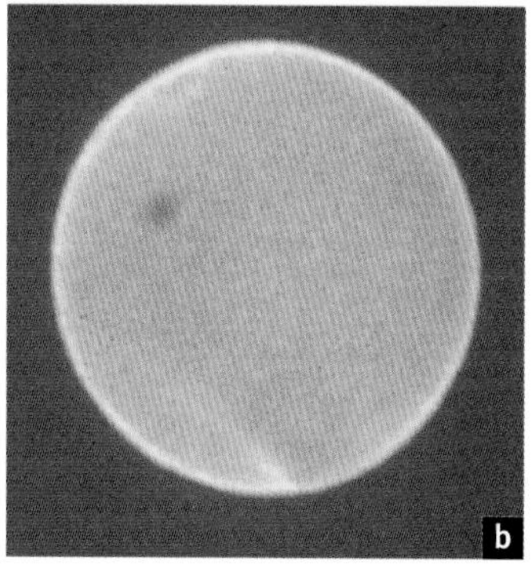

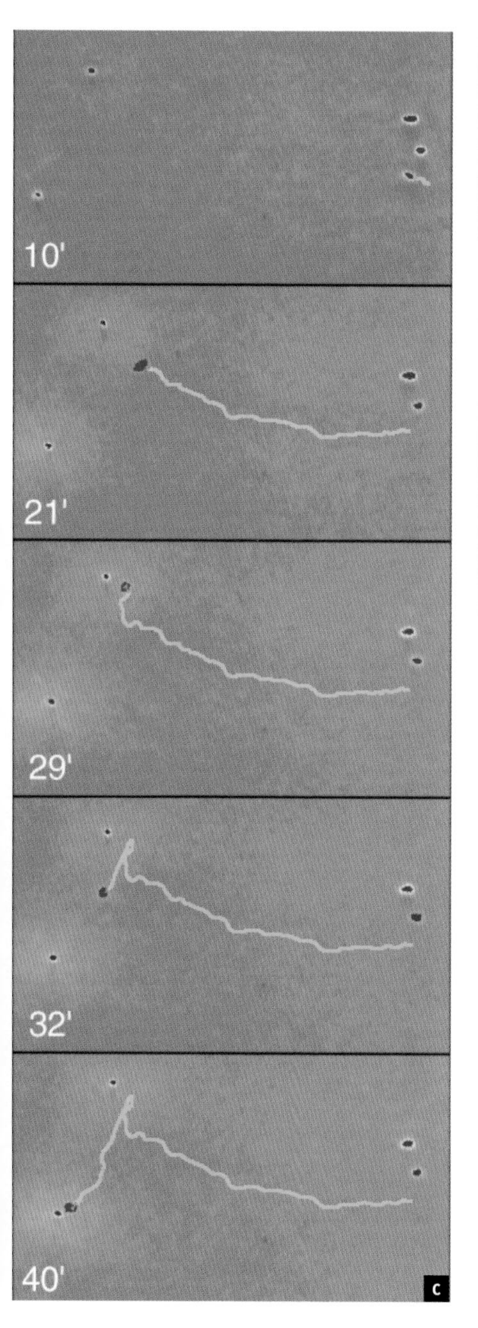

5 Momentous encounters — the bizarre interactions between sperm and egg in a comb jelly *Beroe ovata* (chapter 6): (a) an adult comb jelly; (b) a single egg in which a remarkable form of female choice takes place; (c) the sequence of events occurring after two sperm have penetrated the egg. The nucleus of each sperm, lying on the left-hand side of the picture, is stained blue. The female nucleus is the lower of the three red dots (the other two are the polar bodies produced after meiosis). At time zero the female nucleus has just started to move towards the upper sperm; after checking out this one, she turns to visit and fuse with the other to form an embryo. Numbers indicate time in minutes. See chapter 7.

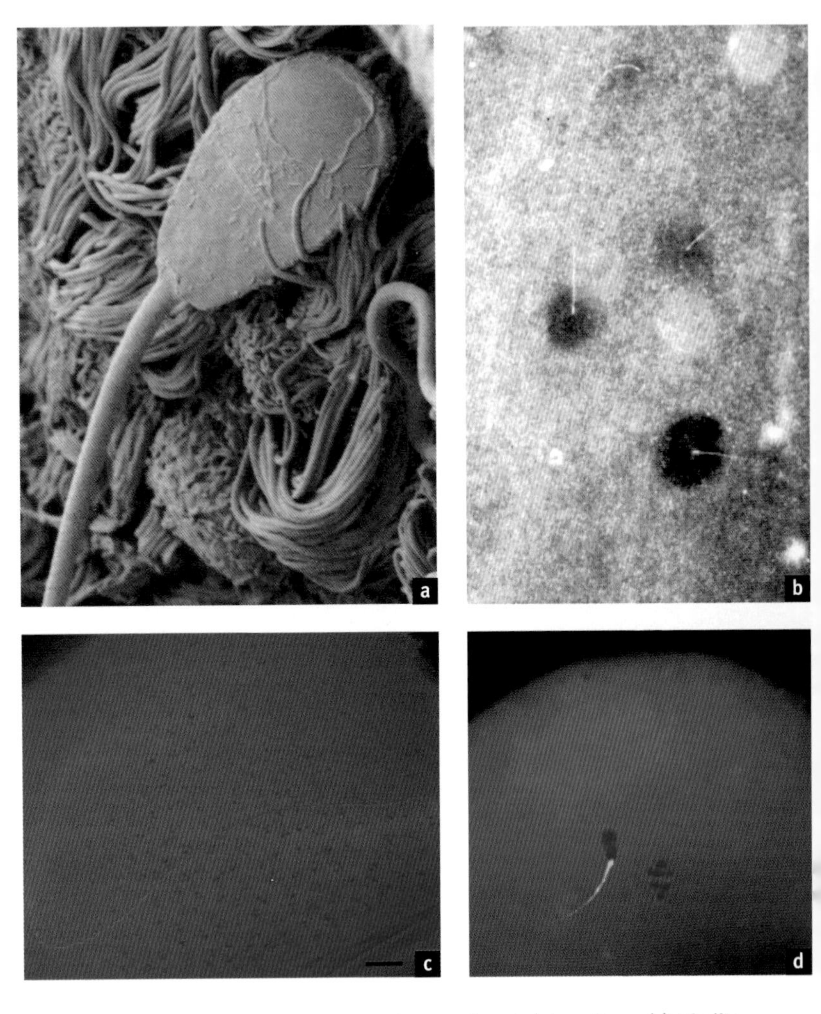

6 More encounters: sperm–egg and sperm–female interactions. (a) A bull's sperm clasped by projections from the cow's oviduct which help transport it towards the point of fertilization (chapter 4). (b) The egg of a bird (a long-tailed tit) showing how four sperm have digested holes in its surface, one of which will pass through the hole to fuse with the female nucleus (chapter 6). (c) A single, gigantic fruit-fly sperm inside the egg (chapter 4). (d) A human egg with a single sperm attached to its exterior moments before the male and female genetic material fuse. The sperm has just penetrated the egg — the purple shapes to its right are the female chromosomes (chapter 5).

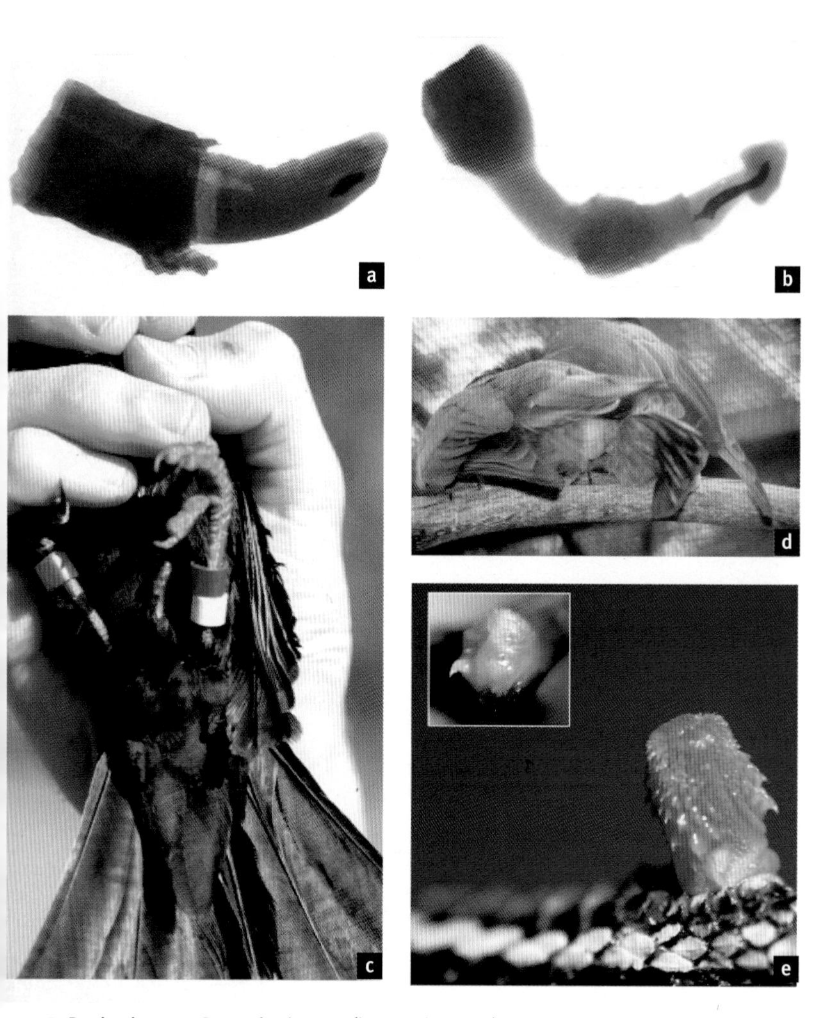

7 Penis pictures. Several primates (but not humans) possess a penis bone or baculum, revealed here by X-rays: (a) a chimpanzee and (b) a vervet monkey (chapter 3). (c) A bird with a false pecker: the red-billed buffalo weaver's pseudo-penis is unique (chapter 3). (d) Also unique, the Vasa parrot uses its huge phallus-like cloaca to acheive a copulatory tie (chapter 5). (e) A tool within a tool: the prickly penis of a red-sided garter snake. This is in fact one of the snake's two penises; at the base of each one there is a single large hook (inset) used to remove the copulatory plug from previously inseminated females (chapter 3).

8 Copulatory competition and conflict (chapter 5). (a) Treading foam : a single female frog is mounted by numerous suitors all competing to fertilize the eggs in her foam nest. (b) Ornithological rape: a forced extra-pair copulation attempt by two male mallard ducks on a single female. (c) Protecting his paternity: a male yellow dungfly remains mounted, even after copulation has ended, to prevent others from inseminating his partner. (d) Traumatic insemination: the male bed bug inseminates his partner directly through her body wall. (e) Complete and utter female control: female spotted hyenas make males grovel for copulations. The male's awkward posture is a consequence of his having to penetrate the female's forward-pointing clitoris. (f) Penis-fencing in marine worms: individuals are both sexes – hermaphrodites. Each rears up and with its everted penis attempts to inseminate the other without being inseminated itself.

morphology was nevertheless substantial. Remarkably, Retzius did not start to look at sperm until he was sixty, but in the years that followed he described the sperm of over four hundred different species of mammal, bird and invertebrate. His wife's fortune enabled him to procure his own specimens, including several great apes. Indeed, Retzius was the first to record the remarkable intraspecific variation in sperm morphology in the gorilla and comment on how similar this species was to ourselves in this respect.

What Retzius's microscopy revealed was that, although the sperm of most organisms, including ourselves, are tadpole-like, incredible variation exists in the size and shape of different species' sperm. Proturans, the most primitive of insects, for example, have disc-shaped sperm which are quite incapable of independent movement. Some millipedes also produce immotile, but crescent-shaped, sperm. The sperm of whirligig beetles occur in aggregations which under the microscope look like feathers: hundreds of individual sperm attached to a central rod. Sperm aggregations of different size occur in a range of insects. At one extreme sperm may occur as an entwined couple as in silverfish, or they may form larger groupings as in grasshoppers and mantis, in which hundreds or thousands of sperm are joined at the head to produce Dali-esque spermatodesms. Elsewhere in the animal kingdom sperm aggregates are rare, but in American marsupials, such as the opossum, sperm operate like well-coordinated Siamese twins. Joined at the head, pairing improves their swimming performance. It works like this: the flagella beat in equal but opposite synchrony with the result that the usual side-to-side movement of the sperm head is offset by the beating of the other sperm. The result is less lateral movement and more forward progression. One consequence of this improved locomotory performance is an increase in the proportion of sperm making it to the vicinity of the eggs: about one for every twenty sperm inseminated. This is much more efficient than, say, a rabbit, where only one in every 5000–10,000 sperm gets anywhere near the eggs. The opossum's efficiency in turn may have enabled it to cut down on the number of sperm males need to produce, store and ejaculate. Compared with other marsupials of similar size, the opossum stores and ejaculates relatively few sperm. In case you were wondering, the paired sperm separate just before they encounter the egg and only one of them fuses with the female pronucleus.[18]

In Any Shape, Size or Form ...

Although sperm vary considerably in the shape of the head and the acrosome – rounded in humans, pointed in finches and sickle-shaped in rats – the most obvious source of variation is the length of the tail or flagellum. In different fruitfly species sperm vary in length from one-third of a millimetre (300μm) to almost six centimetres (60,000μm).[19] Mammalian sperm length ranges from a mere 28μm in the porcupine to 349μm in the honey possum – a twelvefold difference.[20] The sperm of fish also vary considerably, from amoeboid, flagella-less cells in the elephant fish to more conventionally shaped 272μm-long sperm of the Australian lungfish. Similar variation occurs within all animal groups. The question is why?

There are three main ideas for why the sperm of different species differ so much in size and shape. The first has nothing to do with sperm competition and assumes that differences have evolved through natural selection to prevent fertilization occurring between members of different species. The other two ideas depend on sperm competition, and propose that differences in sperm size have evolved as a result of sexual selection, either through female choice of sperm, or male–male competition – competition between the sperm of different males for fertilization.[21]

The first idea is part of a package of tricks to prevent individuals fertilizing the wrong species – so-called reproductive isolating mechanisms. The argument is similar to the lock-and-key idea for genitalia, discussed and rejected in the last chapter. We can probably also reject the idea that differences in sperm evolved to minimize the risk of fertilizing the wrong species' eggs. Although I am not aware of any direct evidence, the mechanisms that evolve to prevent the wrong species fertilizing each other are more likely to operate at earlier rather than later stages in reproduction. For example, it is more efficient all round that as humans we recognize and reject bonobos as potential copulation partners rather than rely on our gametes to put us right.

The second and third ideas assume that bigger means better and that larger sperm are able to out-compete those of other males. Across animal groups as diverse as primates, birds, insects and nematodes, larger sperm are associated with high levels of sperm competition. Species in which sperm competition is most intense tend to have larger sperm than those in which sperm competition is rare or absent. There is even some evidence from within species that larger sperm are more successful. The sperm of the bulb mite are amoeba-like and have no flagellum, but in competition with other males sperm size is an important factor determining fertilization success. In addition, male mites that produced large sperm also produced a more consistently sized product.

What makes larger sperm more successful? The first idea is that females prefer larger sperm and by doing so their reproductive success is increased. Larger sperm may provide (a) direct benefits, (b) genes for 'attractiveness', or (c) genes for viability. Let us take these in turn. The most obvious direct benefit females could acquire from large sperm is some kind of nutrient contribution to their fertilized ovum – the zygote. This idea has been tested in a remarkable comparative study of fruitflies whose sperm vary dramatically in length. Scott Pitnick at the University of Syracuse has spent most of his scientific career trying to understand why this variation exists. He is particularly interested in giant sperm, and especially those of *Drosophila bifurca*. From head to tail this fly measures just 1.5mm, but its sperm are 58mm in length. That's about as long as your little finger. The transfer of a 58mm-long sperm from male to female is potentially messy and unwieldy. The fly's solution is to roll the sperm up and deliver them one at a time. Visible with the naked eye, a single sperm is little more than a pinprick, but under the microscope it looks like a tangled ball of string. The first task for Pitnick and his collaborator Tim Karr was to measure them. But how? Conventional tactics were useless. They opted for virtual reality. Using numerous two-dimensional pictures of untangled sperm, Pitnick and Karr generated a truly gigantic, three-dimensional virtual image they could get inside to track and measure the sperm's numerous coils.[23] The question Pitnick and Karr have tried to answer is whether the long tail of a fruitfly's sperm contributes in some way to the zygote. To do this they compared what happens at fertilization in a number of different species whose sperm vary in length. Their prediction was that if longer sperm have evolved to provision the zygote, a greater proportion of the tail of the longer sperm should be retained inside the egg at fertilization. But the opposite was true. In *D. bifurca* only 1.6mm of its 58mm tail was kept within the egg, whereas in *D. melanogaster* 1.8mm of its 1.9mm sperm were retained. The role of the retained portion of tail remains obscure and its end is ignominious – it ends up inside the gut of the fly larva and is defecated out as soon as the larva hatches from the egg.

The second reason why a female might preferentially utilize long sperm over shorter sperm is that by doing so she gains a genetic benefit for her offspring. Long sperm, or sperm of a particular shape, may be more 'attractive' to the female or, more specifically, to her reproductive tract than others. By having her eggs fertilized by attractive sperm the female will have sons that in turn will produce attractive sperm and daughters that will prefer to be fertilized by attractive sperm. In other words, sperm 'attractiveness' is heritable, but the features that make the sperm attractive have no inherent value; they simply stimulate the female tract in a way that means it is more likely to be used for fertilization. This idea has not been tested, but having eliminated the zygote-provisioning idea, Pitnick thinks this to be the most likely explanation for *D. bifurca*'s giant sperm. These huge sperm take seventeen days to manufacture, considerably longer than the smaller sperm of other fruitflies. The delay arises because the male has to grow enormous testes in which to produce his giant sperm. This in turn suggests that the benefits of giant sperm must be considerable to compensate both for the delay in the onset of reproduction and the reduced number of sperm. Although Pitnick considers 'attractiveness' to be the most plausible explanation for these giant sperm, there is another possibility.

The third type of benefit arising from female choice of sperm is that, rather than the 'attractiveness' of sperm being random, as in the last example, it accurately reflects the viability of the sperm's producer. Put another way, a female can gauge the quality of a male through the attractiveness of his sperm, and by having her eggs fertilized by 'attractive' sperm she produces high-quality offspring *and* sons whose sperm are likely to be successful.[24] Although there is good evidence that the competitive ability of sperm is heritable and while it seems logical that high-quality males should produce high-quality sperm, I know of no studies that have demonstrated that individuals fathered by competitively successful sperm are any more viable than others. A fascinating and unexpected aspect of sperm success is the role of the male's mother. The fertilizing capacity of sperm depends to a large extent on their mitochondria – their energy source. The mitochondria also contain their own DNA and in most organisms the mitochondria are passed directly from the female to her offspring in the fertilized egg. The ability, therefore, of a male's sperm to compete effectively for fertilization is ultimately controlled by his mother!

The final idea that might explain the variation between species in sperm size is male–male competition. The basic idea is that larger sperm are more competitive and more likely to fertilize a female's eggs than small sperm. The success of large sperm may result from their (a) being faster and getting to the egg before their competitors, (b) having greater energy

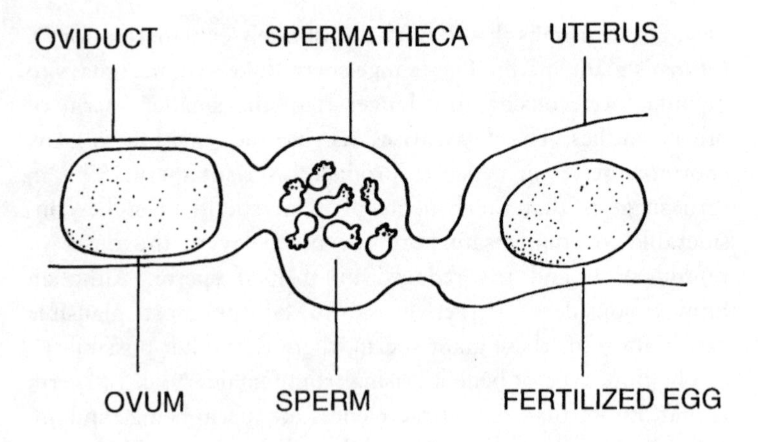

Figure 9

The female reproductive tract of the nematode worm *Caenorhabditis elegans* – the ova move from left to right, fertilized by sperm in the spermatheca (from Ward and Carrel (1979)). See also Plate 2 for a picture of the sperm.

reserves, and hence being more competitive because they survive for longer in the female reproductive tract, and (c) being more efficient at evading female counter-adaptations.

We can address each of these sub-hypotheses in turn. The first idea, and this is one initially proposed for mammals, is that longer sperm move faster than shorter sperm. The underlying assumption here is that the competition between sperm for fertilization is a race and the male with fastest sperm will win. The results from a study of worms provides convincing support for this idea. The sperm of nematodes are amoeba-like, and Craig LaMunyon at the University of Arizona has shown that larger sperm crawl faster than slower ones and fertilize more eggs.[25] The way this works is remarkable: every time the female produces an egg it passes through her sperm store, where it is fertilized, but as it does so the egg flushes out all the other stored sperm (figure 9). Once the egg has passed through the oviduct, the sperm race to get back inside the sperm store, ready to fertilize the next egg, and the larger, faster sperm win. In other species with more conventionally shaped sperm – those with a flagellum – the idea that sperm length dictates velocity has not stood up to scrutiny. The speed with which sperm swim undoubtedly affects their chances of fertilization, but there is no clear indication that longer sperm swim any faster than shorter sperm. This is what theoreticians predict: although a longer tail can generate a greater force, this is completely offset by the increased drag it generates. All else being equal, what this means is that the swimming speed of a sperm should be independent of its length. Direct measurements of the velocity of swimming sperm seem to bear this out: longer sperm do not swim any faster than shorter sperm in either birds or mammals. If size doesn't equal speed, what advantage does a bigger sperm have? The second possibility is that longer sperm survive better than shorter sperm. There is no evidence either way on this point.

The third potential advantage of long sperm is that it enables males to overcome female counter-adaptations. A striking example is provided by a comparative study of birds. The sperm of different songbird species vary in length from about 50μm to 300μm, and species in which sperm competition is rare or absent have shorter sperm than those experiencing intense sperm competition. There is no evidence that longer bird sperm swim faster or survive longer, but one pattern is very clear: the length of sperm in different species closely matches the length of the female's sperm storage tubules (figure 10). This suggests that sperm and tubules have co-evolved and implies that females may have been the main target of selection generated by sperm competition.[26] Jim Briskie, now at the University of Canterbury in New Zealand and Bob Montgomerie at Queens' University, Ontario, proposed that as the level of sperm competition increased, selection favoured females with longer sperm storage tubules because (in some unspecified way) this enabled them to retain control over the way sperm were utilized.[27] The co-evolutionary response of males was to produce longer and

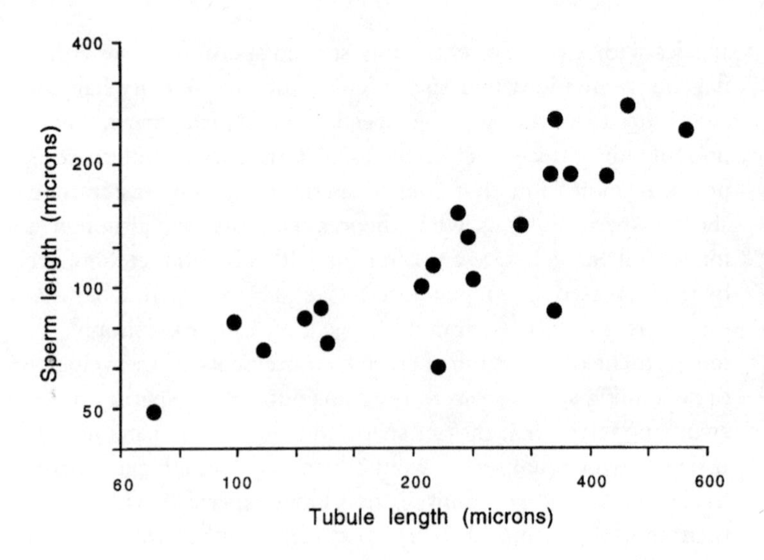

Figure 10

The relationship between the total length of sperm and the length of the sperm storage tubules in passerine birds – each point represents a different species (from Briskie and Montgomerie (1993)).

longer sperm in an ongoing race to retain some control over fertilization.

Before concluding this section, there is one final point. If larger sperm outcompete smaller sperm, selection will favour males with larger sperm. Continuous directional selection for larger sperm should increase sperm size and decrease the variability in sperm size among members of a population. It is quite feasible for selection to produce such a change since we know that sperm size is heritable. What is more, artificial selection experiments have produced changes in sperm dimensions in just a few generations.[28] However, despite what appear to be clear advantages to large sperm, considerable variation in sperm size still persists – even in those animal groups, such as birds, in which sperm competition seems to favour larger sperm. The question then is: what puts the brakes on sperm evolution? The answer is probably the trade off between sperm size and numbers. If both the size and number of sperm influence the outcome of sperm competition, then the optimal balance between these two attributes might vary markedly between different species.

Individual Variation within Ejaculates

Sperm can also show a lot of variation *within* a single individual. The North American fruitfly *Drosophila pseudo-obscura* produces two sizes: long (224μm) and short (120μm), but only the longer sperm morphs are involved in fertilization.[29] The shorter morph contains a nucleus and appears to be fully equipped for fertilization, but does not engage in fertilization. This raises the question of what the shorter sperm are for. The same question has been asked ever since 1905 when biologists discovered that some sperm in the ejaculates of butterflies had no nucleus at all and so are clearly incapable of fertilization. In some species these non-fertilizing, non-nucleate sperm* make up over 90 per cent of the ejaculate, so it seems likely that their role is important. There have been three main ideas for the function of non-nucleate sperm. First, perhaps they help to transport the nucleate sperm from the testis through the female tract to the point of fertilization. Second, they may provide some form of nourishment to the nucleate sperm or even to the female or to the zygote. There is no evidence for either of these suggestions. The third idea is that non-nucleate sperm serve a role in sperm competition. This could work in one of two ways: non-nucleate sperm could either help to reduce the effectiveness of sperm from previous inseminations, or they might be a cheap way of making a bulky ejaculate – if the female feels that her sperm store is full she is less likely to copulate with another

* Technical terms: apyrene = non-nucleate; eupyrene = nucleated, fertilizing sperm.

male. A recent study of a moth, confusingly referred to as the army worm (because its larvae march around *en masse* in search of food) showed that in a typical copulation a male inseminates 240,000 sperm, 95 per cent of which are non-nucleate. It was the number of non-nucleate sperm remaining in the female's sperm store, and not the number of nucleate sperm, that determined when she remated. These results provide the best evidence so far for the cheap-filler hypothesis – the idea that non-nucleate sperm are fillers, whose role is to fool the female into thinking she has more fertilizing sperm than she really has.[30]

Sperm Profligacy?

When it was suggested, in the eighteenth century, that only one of the millions of sperm inseminated was necessary for fertilization, God's wisdom was called into question. When it was confirmed that only one or a few sperm in an ejaculate achieve fertilization the efficacy of natural selection was called into question. For an evolutionary biologist sperm profligacy is counter-intuitive. Any male that inseminates more sperm than is necessary to secure a fertilization should be penalized by natural selection. The question is: what *is* the evolutionary advantage to inseminating so many tiny sperm? With the discovery of sperm competition, Geoff Parker provided the answer, or at least part of the answer (see chapter 2). When more than one male copulates with a female, inseminating more sperm is like buying more lottery tickets: more tickets means a greater chance of winning and more sperm mean a greater chance of fertilization.[31] As we saw in the last chapter, species in which sperm competition is rife invest heavily in sperm-producing tissues since big testes produce more sperm.

Sperm almost always outnumber eggs, but the ratio of sperm presented to eggs varies markedly across species. In some species of insects there are less than a hundred sperm for every egg but, at the other extreme, there are hundreds of millions of sperm for each egg – as in humans. In the 1960s, in the days before sperm competition was even a twinkle in Geoff Parker's eye, Jack Cohen came up with a solution for why males should inseminate so many sperm.[32] Cohen suggested that sperm profligacy resulted from production errors. Despite thinking that sperm are cheap to produce (see chapter 3), his argument was that perfect sperm are difficult to make. All that DNA has to be very carefully condensed and packed into the sperm head and, as in any mass-production system, errors are common. If it is the absolute number of perfect products that counts, in a faulty system the simplest solution is to produce more. Cohen supported his idea by finding a significant correlation between the degree of sperm redundancy and a particular production event during the manufacture of sperm. The production event is the crossing over of the meiotic chromosomes during the formation of sperm. In different species different numbers of crossing-over events are needed to produce sperm, from none in the honey bee to about fifty in the goat. With each crossing over there is a 20 per cent chance that something will go wrong and produce a faulty sperm. In humans there are fifty crossovers per sperm so the chance of an error occurring is high, so much so that Cohen estimated that only one sperm in a million might be suitable for fertilization. An important corollary of this production-error hypothesis is that there would have been extremely strong selection on females to avoid being fertilized by defective sperm. That is, following insemination females must have a very effective system for weeding out the defectives. As we saw in chapter 3, this is exactly what occurs.

Cohen's explanation for the correlation between sperm redundancy and crossover frequency was extremely clever, but subsequent research revealed an inconsistency. If Cohen's interpretation of the correlation was correct it should apply equally to female sex cells, but a follow-up study revealed no correlation whatsoever between egg redundancy and crossover frequency. More recently still, John Manning and Andrew Chamberlain in Liverpool came up with an even more ingenious solution, which elegantly combined Cohen's findings with sperm competition theory.[33] Cohen assumed that crossover frequency was fixed for particular species which then produced an appropriate number of sperm to compensate for the problem of production errors. Manning and Chamberlain turned this idea on its head and said that the more crossing over there was the less likely it was that production errors would occur. After all, the crossing over mixes up the genetic material between chromosomes and *reduces* the chances of getting bad combinations of genes. Their argument was that, if mutations arise in sperm that reduce their competitiveness, there will be competition between sperm from the *same* ejaculate to fertilize a female's eggs. If – and this is the important point – there is also competition between the sperm from *different* ejaculates (i.e. regular sperm competition) then there will be strong selection for both increased sperm numbers *and* increased levels of crossover, because the male whose ejaculate contains the greatest number of non-defective sperm will fertilize most eggs.

Male Control

It is easy to see how sperm competition would have favoured relatively large ejaculates. In species where females routinely copulate with more than one male a male's reproductive success will tend to be enhanced the more sperm he transfers. In other species where sperm competition is less intense, selection has favoured males with more modest ejaculates. Although each species has a typical number of sperm per ejaculate (see table 1 in chapter 3), as anyone who has ever done any sperm research knows, the numbers of sperm vary enormously between ejaculates even in the same individual. This variation, much of which appears to be due to chance, is a constant source of irritation to researchers since it renders their results so variable and difficult to interpret. Some of the variation in sperm numbers arises from a remarkable male facility.

For any particular species the likelihood or intensity of sperm competition varies from situation to situation. Imagine a species in which females use some sperm to fertilize their eggs soon after insemination. A male encountering a virgin female has relatively little risk of sperm competition compared with a male encountering a female who has already been inseminated. To achieve fertilization the optimal ejaculate size differs in these two different situations. With a virgin female fewer sperm might do the job, but with a non-virgin a male would do better by inseminating a much larger number of sperm. Of course, to be able to adjust his ejaculate size appropriately, a male must possess both the necessary anatomical architecture and the wherewithal to distinguish between virgin and non-virgin females. There is now abundant evidence that males of some species possess both these attributes. Paul Jivoff studied the reproduction of blue crabs for his doctorate at the University of Maryland and found that males were very sensitive to the insemination status of females.[34] If a male bumped into one of the 12 per cent or so of females that had copulated with another male, he inseminated many more sperm than if he met a virgin. How do males know the difference between virgin and non-virgin females? In the blue crab's case, the male may be able to feel whether a female has been previously inseminated by the use of a pair of elongated appendages, technically known as pleiopods, which are used to transfer sperm into the female. These elongated structures are placed inside the female, and can apparently detect whether a female's sperm store contains sperm or not. If it does, the male copulates for longer and inseminates more sperm.

The most dramatic examples of ejaculate adjustment come from externally fertilizing fish. The blueheaded wrasse is a tropical reef fish with an unusual breeding system and intense sperm competition. Females are attracted to the large, brightly coloured males who defend traditional spawning sites. Once they have secured a territory a male keeps it for his entire breeding life – just three months, after which he dies. This species has an unconventional lifestyle since males may start life as either a male and grow up to become a dominant individual, or they may start off as a female, and only when they have attained a good size change sex and start to operate as a male. The pay-offs from being dominant breeding males are enormous, with dozens of females turning up during the afternoon spawning session each day. Consequently, territorial males have been under strong selection to produce copious numbers of sperm. At the simplest level, with no other males around and no risk of sperm competition, a territorial male judges how many sperm to release simply on the basis of the female's size: the number of eggs she releases corresponds closely to her size. In an ideal world a male would release sufficient sperm to fertilize all the eggs a female releases, but he does not do this – because his interests and those of the female do not coincide. In an evolutionary sense each female wrasse wants *all* her eggs fertilized. The male, on the other hand, is not particularly bothered about ensuring this. Instead, his aim is to maximize his overall reproductive success and he does this best by selling females short. He ejaculates slightly fewer sperm than are necessary to fertilize all their eggs, but by doing so is able to fertilize the eggs of a greater number of females in total – a clear example of sexual conflict, and one that males appear to win.[35]

There's more. In the presence of other males – sneaky spawners that zip in and ejaculate at the same time as the territorial male wrasse – things become more complicated. With a moderate number of sneakers the territorial male simply ejaculates more sperm. But if the number of sneakers increases beyond a certain threshold he reduces the number of sperm he releases.[36] A male's ejaculation strategy, the amount of sperm a male is prepared to allocate to a particular spawning event, is analogous to how much effort you might be prepared to expend defending your reserved seat on a train. If one or even two other passengers contest your seat, you are prepared to spend time and effort arguing it out, but if a bunch of football hooligans decide they want your seat, your best strategy (in terms of your long-term fitness) is to acquiesce. When the competition is intense and the chances of fertilization low, there's no point wasting sperm.

The precise adjustments male wrasse make to their ejaculates involve integrating information on both the size of the spawning female and the strength of the competition. This is the kind of biological sophistication that creationists revel in, pointing out how unlikely it is that such precise control could evolve by chance alone. For those with an appreciation of evolution, the wrasse's ejaculatory precision merely demonstrates the effectiveness of the evolutionary process. The wrasse does not make a conscious decision about how many sperm to release. Rather, natural selection has eliminated those males that failed to adjust their sperm numbers in an appropriate manner.

Seminal Magic

A mammalian sperm can do three tricks: it can swim; it can swim very fast; and it can eat a hole in the outer covering of an egg. Regular swimming is achieved by the flagellum, which, powered by the energy from the mitochondria, beats in a characteristic fashion. The sperm of different mammal species swim in a particular manner and at a particular speed, presumably in response to the specific features of the female oviduct they find themselves in. If you examine a sample of live sperm under a microscope the overwhelming impression is one of movement: fast, furious and erratic. A common misconception is that it is this vigorous swimming that gets the sperm from the point of insemination to the ovum at the other end of the oviduct – in humans a distance of 15–20cm. But this is less than half the story. In reality sperm move by a combination of their own motility and active transport by the female tract: rather like walking on an airport moving pavement – but not quite as simple. Studied extensively by reproductive physiologists, sperm transport turns out to be mind-bogglingly complicated, involving a sophisticated series of interactions between the sperm, seminal fluid and the female tract. For example, the seminal fluid in which the sperm are delivered to the female affects both the way sperm move and the way the female tract responds to the sperm. Moreover, the outer surface of the sperm themselves changes as they move along the escalator. Far from being a direct run, as well as altering its speed along its length, the escalator contains numerous traps and filters, so that few of the initial recruits ever make it to the end.[37]

As the sperm ascend the oviduct they are prepared for fertilization in a process called capacitation. For reasons unknown, capacitation appears to be confined to mammals; the sperm of reptiles and birds have no need to ready themselves in this way. For most of their journey mammalian sperm remain in close physical contact with the surface of the female reproductive tract – as though they are stuck to the escalator. When the sperm get near the top of the tract, in the isthmus, the process of capacitation begins. And at the same time, and possibly in response to ovulation, the sperm undergo a dramatic change in the way they behave. Instead of moving in a steady, more or less forward manner, they become hyperactive. Their faster and more erratic movements help to wrench them free from the surface of the oviduct, thereby maximizing their chances of bumping into the ovum. Having done that, their extra efforts may also play an important role in penetrating the ovum.[38]

What determines which sperm from an ejaculate will fertilize a female's eggs? This is the question to which andrologists and those working in *in vitro* fertilization clinics would most dearly like an answer. Despite years of research and huge financial incentives they do not know the answer. What andrologists do know with some certainty is that ejaculates from different males vary considerably in their fertilizing ability. In some cases the reason is obvious: an ejaculate with relatively few sperm, relatively few normal-looking sperm or sluggishly swimming sperm is less likely to result in fertilization than one in which sperm are abundant, sleek and vigorous. However, among non-human animals, males whose sperm appear completely normal can show big differences in their fertilizing capacity, with some individuals consistently outperforming others. My colleague Harry Moore in the University of Sheffield has looked at what makes a great pig ejaculate. Motility is the answer – not simply how fast, but a combination of how fast and for how long. Pigs and, it turns out, people who produce sperm that stay fast for a long time have the best chances of fertilizing.[39] The same is true for the domestic fowl: cockerels with vigorous sperm are the best fertilizers.[40] The frustrating thing from my point of view was that until recently all these experiments have been conducted using the sperm of only a single male. What was really needed were sperm competition experiments in which equal numbers of live sperm differing in their mobility from two different males were mixed and inseminated into females. I collaborated with David Froman at Oregon State University to conduct this experiment using cockerels that had previously auditioned for their sperm mobility parts. When we pitted fast- and slow-sperm males against each other the fast-sperm males won, hands down.[41]

A Commotion in the Semen

It was once assumed that the plasma portion of the blood was merely a vehicle for the red and white cells. But then Louis Pasteur and others showed that lurking invisibly in the serum lay the immense complexity of our immune system. In much the same way, the seminal fluid was originally viewed as a vehicle for the animalculae or sperm animals. However, we should have had some inkling that there was more to seminal fluid than meets the eye since its volume and consistency vary so much between species. The male turkey, endearingly known as the 'tom' in the poultry business, produces an ejaculate with virtually no seminal fluid: a mere 0.22 microlitres of semen containing 1,600 million sperm. At the other extreme, the male pig produces half a litre of semen containing 100,000 million sperm. To put this another way: each pig sperm wallows around in 5 microlitres of seminal fluid, whereas each turkey sperm has a mere 0.0001 microlitres in which to swim.[42]

Mammalian reproductive physiologists have documented in great detail the chemical composition of seminal fluid in a range of domestic species including the bull, the boar and the ram. Their interpretation of the huge variation in composition has been rather limited and, if quizzed, they are content to stick with the idea that seminal fluids provide a protective and nurturing environment for the sperm.[43] It was not until recently that a more cynical possibility was proposed. As well as protecting sperm, seminal fluid may play a role in manipulating the female. If you think about it this is an obvious extension: any male that produced substances in his seminal fluid that facilitated the transport of his sperm in the female's oviduct, or speeded up the rate at which the female ovulated, or did anything at all that increased his chances of fertilization would be at a huge competitive advantage in sperm competition.

The fact that this is exactly what happens is beautifully exemplified by the most unlikely of organisms: the common house fly.[44] The house fly's seminal fluid is a veritable cocktail of chemicals. Some of these mimic substances that occur naturally in the female's own body, triggering her egg-laying processes. By inducing a female to lay eggs sooner than she might otherwise have done, the male house fly makes the female use his sperm to fertilize her eggs. It is straightforward to see how this evolved. Imagine a mutant whose seminal fluid contains a strong persuader. Compared with a male that didn't trick females into premature egg-laying, all else being equal, the persuader male would leave more offspring simply because the female would be likely to use his sperm before she died. Once females started to copulate with several males, for whatever reason, the benefit to a male of accelerating a female's egg-laying would be even greater.

The second type of compound that occurs in a house fly's seminal fluid is one that puts the female off sex. Such anti-aphrodisiac substances occur in a number of insects and snakes.

In some species they merely delay a female's return to receptivity, but in the house fly's case the effect is dramatic: it puts the female off sex for the rest of her life – ensuring, as a consequence, that the male fertilizes all her eggs. I'm always staggered that an organism that we consider so inconsequential and swat with such impunity should have such remarkable processes going on inside its tiny body.

The ultimate ejaculate is one that can kill or disable the sperm of rival males already stored in the female's reproductive tract. As we saw with Bateman's fruitfly experiments, in many (but not all) species, the last male to copulate with a female fertilizes most of her eggs. It has also been known for some time that the more often a female fruitfly copulated the less time she lived. Initially this reduction in female longevity was assumed to be because bearing eggs was costly. Since males suffered no similar reduction in lifespan as a result of frequent sex it was further assumed that turning out eggs was costlier than turning out sperm. However, a series of experiments revealed that it was being inseminated rather than producing eggs that caused the reduction in female lifespan. And then, rather like Spallanzani trying to discover which component of semen resulted in fertilization, researchers designed experiments to establish whether it was the sperm themselves or the seminal fluid that was so damaging to females. Luckily, the researchers had access to a spermless mutant: a fruitfly that inseminated seminal fluid but no sperm. Females who copulated with these males showed exactly the same reduction in lifespan.[45]

Then the penny dropped. Last-male sperm precedence and a reduced female lifespan were linked. Just as in the house fly, the male fruitfly transfers female-manipulating chemicals in his semen – speeding up oviposition, dissuading females from copulating for a while and, most importantly, disabling the sperm of rival males. However, the substances responsible for incapacitating the sperm of other males are chemically similar to those in spider venom and are toxic to the female fruitfly. From the male's point of view this is irrelevant: as long as she lays a batch of eggs fertilized by his sperm any reduction in her lifespan is of no consequence. In the case of *Drosophila* it appears that this particular battle of the sexes is won by males.[46]

You might be asking why females allow themselves to be manipulated in this way. The answer is that it may be difficult for them to evolve counter-measures, particularly if male seminal substances mimic the female's own hormones which stimulate egg-laying. In addition, female counter-measures will evolve only if it is worthwhile for the female. That is, if the cost of doing so in terms of time or energy does not outweigh the advantage. However, the co-evolution of male and female reproductive traits in *Drosophila* is an evolutionary arms race every bit as ferocious as those between predators and prey and between parasites and their hosts. In an ingenious experiment William Rice at the University of California at Santa Cruz showed that when female fruitflies were prevented from co-evolving with males, the male's sperm displacement abilities rapidly evolved to become even more damaging to females.[47] The implication of this experiment is that, under normal conditions, females are evolving as fast as they can in response to male attributes, and lagging just slightly behind. In another study, Rice and his graduate student, Brett Holland, created the opposite situation and removed the opportunity for sexual selection by forcing male and female fruitflies to be sexually monogamous. In stark contrast to the situation in the wild and in the previous experiment, the interests of each sex were now very similar. As predicted, over the course of several generations, ejaculates became less and less dangerous to females – since under these circumstances it was very much in the males' interest to keep females reproducing for as long as possible.[48]

This chapter has explored the main features of male and female sex cells and, for males, the media in which sperm are transported into the female, in terms of sperm competition. The emphasis has been predominantly on males, not because of chauvinism, but because so much more is known about ejaculates than about ova. This is partly because ejaculates are inherently easier to study than ova, although I cannot entirely exclude the possibility that this is a chauvinistic perception. The greater variation in ejaculates compared with ova is also consistent with the view, introduced in the first chapter, that selection operates more intensively on males than on females. In the next chapter we look at how sperm and egg from the two sexes get together to create new individuals.

5 Copulation, Insemination and Fertilization

Homo sapiens is in peril of copulating itself to extinction.
K. MAXWELL, *A Sexual Odyssey* (1996)

The animals we are most likely to see copulating are insects, house sparrows, dogs and occasionally farm animals. Because it is generally considered bad form, we rarely look closely. Even less often do we consider why some insects should remain in copula for so long, or why house sparrows copulate ten or twenty times in rapid succession, or why dogs form a prolonged copulatory tie capable of resisting any number of buckets of water. In most cases the purpose of copulation is insemination: the transfer of an ejaculate from the male to the female. And in most cases the purpose of copulation and insemination is fertilization, although the chances of any one insemination resulting in fertilization differ enormously between different species. Given the number of sperm in a single ejaculate, a female needs to copulate only once or at most a few times in a single breeding cycle to have sufficient sperm to fertilize all her eggs. The fire ant, an unwelcome introduction to the southern United States, provides the ultimate example. A virgin queen leaves home on a nuptial flight and is caught by a single male who inseminates her with 7 million sperm. The male is nothing more than a one-shot flying penis and soon dies, but afterwards the female has sufficient sperm to last a lifetime – seven years, in fact – over which time she produces 2.6 million fertilized eggs, each requiring an average of just three sperm, a more efficient use of sperm than almost any other organism. At the other extreme, lionesses copulate several thousand times for each pregnancy.[1]

Since a single dose of semen may often be sufficient to fertilize all a female's ova, a potential conflict exists between the sexes over copulation. For males more is almost always better because it increases their chances of winning at sperm competition. For females, less may be better, but more may be a way of minimizing or resolving the conflict with males. The competition between males to fertilize the eggs of females and the conflict between males and females over fertilization are two main evolutionary forces that have shaped copulation and insemination.

Competition and conflict between the sexes began long ago, however, before copulation existed, and long before penises or vaginas had evolved, so this chapter begins by looking at external fertilization. We then consider internal fertilization: the deposition of semen within the female reproductive tract which is accomplished by the act of copulation. The focus is on how competition between males and conflict between males and females have shaped the form and frequency of copulation behaviour and, ultimately, the process of fertilization itself.

Sperm Broadcast

Aquatic invertebrates, such as starfish, sea urchins, sea anemones, corals, sea cucumbers and some worms engage in what must be among the simplest forms of reproduction. These animals, referred to as broadcast spawners, simply release their gametes into the water. This system may seem wonderfully uncomplicated compared with our own, but its simplicity belies the problems faced by male and female sex cells as they try to locate each other. The main problem is dilution, and is particularly acute in flowing waters when potential partners may never see each other (even if they have eyes, which many of them don't). Even when partners lie in close proximity, local turbulence may send their sperm and eggs spinning out of control and prevent them from fusing together. As a consequence, getting their eggs fertilized is a major problem for females – a situation almost unheard of for species with internal fertilization. That sperm are often limiting is one of the most important factors moulding the reproduction of broadcast spawners. It may seem counter-intuitive then that sperm competition is also important – but under some circumstances it is. In broadcast spawners there is competition between eggs of different females for sperm (limitation) and competition between sperm from different males to fertilize eggs. Together these processes have favoured the production of massive numbers of sex cells by each sex. Of the two, sperm limitation is generally regarded as being the more important, for several reasons. First, it is unusual for *all* of an individual's eggs to be fertilized. Second, the amount of body tissue given over to the production of sex cells is similar in males and females, an unusual situation compared with internal fertilizers. In addition, the investment in reproductive tissue can be enormous: the gonads in some starfish constitute over one-third of their body weight in both males and females. As well as producing large numbers of sex cells, several other adaptations help to increase the chances of fertilization for both sexes. Among mobile animals, such as starfish, worms and sea cucumbers, individuals of both sexes aggregate prior to spawning. Sessile species, like corals and sea anemones, synchronize their spawning, often using environmental cues to do so.[2]

As a schoolboy in the late 1960s I had delighted in Ralph Buchsbaum's book *Animals without Backbones*, first published in 1933, and still worth reading today. The two aspects of the book that left a lasting impression were Buchsbaum's cherubic visage on the back cover and his description of the synchronous spawning of two species of marine worm. First, the boom-and-burst style of reproduction of the Bermuda fireworm. Each month, a few days after the full moon, female worms emerge and swim around in circles, just after sunset, emitting a greenish phosphorescent glow. The smaller males then appear and flash their own lights on and off as they dart towards the females. When the sexes coincide they simultaneously explode and release their sperm and eggs into the sea. Another species, the Palolo worm in the South Pacific, reproduces in a single mass spawning each year but, unlike the fireworm, lives to tell the tale. Gonads develop in the posterior part of the body, and one week after the November full moon this reproductive portion of the Palolo worm breaks off and swims to the surface, releasing sex cells as it does so. The result of this mass spawning, which lasts little more than one hour, is a sea resembling vermicelli soup which turns milky in due course as the sperm and eggs are discharged. To the Samoans this is manna from heaven: 'They scoop them up in buckets and prepare a great feast, gorging themselves just like we do on Christmas day.' Buchsbaum's account of the worm-fest dates from the 1930s, but the first description was made almost a century before when the appearance of English seafarers and the Palolo worm coincided. The festival continues to this day.[3]

Two aspects of the Palolo worm's exotic reproductive biology not discussed by Buchsbaum were sperm competition and sperm limitation. This type of external fertilization is a free-for-all, creating strong selection pressures on males to produce fast-moving sperm and on females to produce eggs that are efficient in securing sperm for fertilization.

The majority of fish are external fertilizers and they show a remarkable diversity of sexual arrangements, ranging from social monogamy through to the enormous group spawning frenzies of herring, cod and grunion. For the males of some species, such as salmon, spawning may be a once-in-a-lifetime experience; for others, like the blueheaded wrasse, it may occur dozens of times each day.[4]

The high mobility of each sex and the fact that the access to fertilizable eggs is highly predictable, albeit short-lived, means that in many externally fertilizing fish, sperm competition is rife. In species such as the capelin where spawning occurs in trios – two males and a female – sperm competition is inevitable. During an expedition along the Labrador coast in 1867 Lieutenant W. A. Chimmo observed and carefully described their behaviour:[5]

The manner of the Capelin spawning is one of the most curious circumstances attending its natural history. The male fishes are somewhat larger than the females, and are provided with a sort of ridge projecting on each side of their backbones, similar to the eaves of a house, in which the female is deficient. The female on approaching the beach to deposit its spawn is attended by two male fishes, who huddle the female between them . . . In this state they run all three together with great swiftness upon the sand, when the males by some imperceptibly inherent power compress the body of the female between their own so as to expel the spawn from an orifice near the tail. Having thus accomplished its delivery the three Capelin separate, and paddling with their whole force through the shallow surf of the beach generally succeed in regaining once more the bosom of the deep.

Internal Fertilization

When males and females are able to get close together before releasing their sex cells the chances of fertilization overall are increased, especially for the females. But in mass-spawning events, the competition between sperm of different males is also intense. Any male who could get even closer or get his sperm inside a female would be at an advantage over other males. Competition between sperm from different males is thought to have been an important factor in the evolution of internal fertilization,[6] which occurs among a number of marine organisms, such as barnacles. Internal insemination is obviously also a prerequisite for life on land, and the way this transition was achieved was through the spermatophore – a pre-packed ejaculate. Copulation itself probably evolved even later. Males of some of the most primitive animals, such as mites, simply dump their spermatophores on the ground, and females search for them. The sequence of evolutionary events leading to internal fertilization then seems to be have been: external fertilization, followed by internal fertilization with external spermatophores, in turn followed by copulation in which the spermatophore was deposited inside the female, and finally copulation and the transfer of free sperm inside the female.

The initial selective force favouring internal fertilization may have been the avoidance of sperm competition, but the end result has been exactly the opposite. Instead of reducing competition between the sperm from different males, each stage of the transition from external to internal fertilization merely created a suite of new opportunities for males to outdo each other. As well as generating new selection pressures on males via sperm competition, internal fertilization generated other selection pressures because sperm now had to survive and remain viable for long periods inside an extremely hostile female tract. It was precisely to prepare their sperm for this dangerous saga that males of internally fertilizing species evolved special sites for adding the crucial finishing touches to their sperm. Externally fertilizing species such as fish and frogs have nothing equivalent to the epididymis of mammals or seminal glomera of birds (see chapter 3) because their sperm have to negotiate nothing more sinister than water, and usually for only a few seconds or minutes.

For females internal fertilization may have been beneficial, at least initially, because it increased their chances of getting their eggs fertilized. But internal fertilization must have been a mixed blessing, generating a maelstrom of conflicting selection pressures. As well as providing a highway for sperm, the female reproductive tract offered parasites and other pathogens a direct route to the gonads and other internal organs. Females would therefore have been under strong evolutionary pressure to keep pathogenic aliens out but to allow sperm aliens in, or at least some of them. As long as males continued to compete for fertilizations by vying to inseminate more sperm than their competitors, females, as we have seen (chapter 3), would also have to manage the numbers of sperm in their reproductive tract carefully. It would also have been crucial for females to avoid being fertilized by sperm from another species. Breeding males are usually highly motivated and often indiscriminate. Ejaculation carries little cost and there has therefore probably been little selection against males copulating with the wrong species. Indeed, selection may have favoured a lack of discrimination among males, since he who hesitates is lost. But, for a female, insemination and fertilization by the wrong sperm is likely to be very costly – an entire batch of nutritious eggs doomed to failure. Selection then would have driven females to be highly selective about the numbers and quality of sperm they retain – a theme we return to in more detail in chapter 6.

Semen Deposition

In most cases the whole point of copulation is the deposition of semen inside the female's reproductive tract. From the male's point of view, the closer he gets his sperm to the female's eggs, the greater his chance of fertilizing them. From the female's perspective, the further away from her eggs she can keep a male's sperm, the greater the degree of control she has over fertilization. This conflict of interests explains much of the variation in reproductive anatomy: males have evolved long penises to place their sperm in the best position, while females have evolved long reproductive tracts to retain some control. For females the best form of control over reproductive events is the possession of a sperm store. The females of most insects have one or more receptacles in which sperm are kept prior to being used to fertilize eggs. A typical arrangement is for the sperm to be deposited initially into a bag-like structure, the bursa copulatrix, prior to sorting out which sperm are to be stored for later. In the tiny cowpea weevil the male inseminates an average of 46,000 sperm, many more than the female actually needs, directly into her bursa. From these the female takes up a sub-sample comprising 6,200 sperm, or about 15 per cent of the original ejaculate, for longer-term storage in her spermatheca.[7] The sperm remaining in the bursa are digested by the female – waste not, want not. Those sperm in the spermatheca are then used as the female requires them, to fertilize eggs, over the seven to ten days of her egg-laying life.

In birds, at least those without a penis, the swift cloaca kiss hides a sophisticated set of movements. The bird's cloaca is the common exit for both the gut and the reproductive tract. As the two cloacae come together, the female everts her oviduct so that it protrudes slightly out of the cloaca. The male deposits his viscous ejaculate on to this, and as the female retracts her oviduct back into its original position the sperm are drawn into the vagina. All this in some species in a fraction of a second.[8]

The site of semen deposition in mammals is well known. This is not surprising given that there are considerable commercial implications for the production of farm animals, such as cattle, sheep and pigs, and considerable clinical interests in our own reproduction. In this respect mice, rats and hamsters have provided cheap models for human reproduction. Two broad patterns of semen deposition exist. In primates, including humans, ruminants (e.g. cattle), rabbits and some rodents, semen is deposited in the vagina. But in horses, dogs and pigs, the penis is able to penetrate beyond the cervix to place the ejaculate directly into the uterus.[9] Unfortunately we do not know enough about the sexual behaviour of the wild ancestors of these species to assess how important sperm competition has been in determining the location of semen deposition.

Insemination Deviations

When we think of insemination, we usually imagine a species like ourselves. In order to deposit sperm in an appropriate location, the male first places his penis inside the female's reproductive tract. For most species this is an accurate scenario, but in some animals insemination is more bizarre, and, possibly to short-circuit any possible female control, males inject their semen directly into some other part of the female's body. Because this usually involves puncturing the female's skin or exoskeleton, this unlikely process goes by the name of traumatic insemination and it was Geoff Parker's Ph.D. supervisor, Howard Hinton (see chapter 1), who in the early 1960s was among the first to try to understand the adaptive significance of this unusual mode of insemination.

The common bedbug inseminates the female directly through her body wall. He does this by means of a sickle-shaped structure, the paramere, located at the tip of his abdomen. When not in copulation mode the paramere is carefully folded away against the body like the closed blade of a penknife. But during copulation it flips outwards to pierce the female's body between the plates of her cuticular armour. Only then does the penis, which lies within a groove on the paramere, start to transfer sperm into the female. This isn't the end of it. What happens next is also very odd. The female has a special cushion-shaped structure, the spermalege, located just under her cuticle, into which the male injects his sperm. The spermalege is little more than a bag of amoeba-like cells, whose job may be to destroy sperm. Those sperm which escape swim out through the spermalege and in through the wall of the female's sperm store. The sperm continue their progress by swimming along the narrow outer lumen of the double tube (a tube within a tube) which connects the sperm store to the ovary. Once inside the ovary the sperm start to fertilize eggs – an unusual occurrence since in most other insects fertilization takes place only when the eggs have left the ovary.[10]

The driving force behind the bedbug's unusual form of insemination is probably a combination of sperm competition and female sperm choice – the differential utilization of sperm by females. By injecting his sperm close to the ovaries, a male might avoid the sperm of male competitors but he might also simultaneously avoid the possibility that the female favours another's sperm over his. On the other hand, once this kind of traumatic insemination occurred and was successful, any males persisting with the tried and trusted traditional technique would soon be eliminated because their sperm would have very little success. The only evolutionarily stable situation would be for all males to become traumatic inseminators. But they would then be under increasing pressure to inseminate more and more accurately. Selection would also operate on females. Only if females were not disadvantaged by it could traumatic insemination evolve and be maintained. Bedbugs are tough little beasts designed to withstand extreme abuse from their hosts and, unlikely as it may seem, females are probably little affected by being inseminated through their abdomen wall. However, they were not evolutionarily passive. In response to the male's dirty trick they have effectively evolved a new reproductive system – the spongy spermalege – under their exoskeleton at the point where males penetrate. This structure appears to stop sperm in their tracks and allows females to regain some control over who fertilizes her eggs.[11]

The most extreme form of traumatic insemination and sperm competition occurs in *Xylocoris*, a close relative of the bedbug. Here, insemination sometimes involves homosexual rape. Following insemination through the body wall the rapist's sperm make their way to the recipient's vas deferens and are used along with the male's own sperm the next time he inseminates a female. In some cases a male may even be inseminated by another male while he is copulating with a female.[12]

An extraordinary form of sperm injection occurs in some marine flatworms. These exquisitely coloured animals can be seen gliding effortlessly across the corals in tropical seas – if you happen to be scuba diving. Despite their gentle manner, their sex lives are potentially damaging. Flatworms, like earthworms and leeches, are usually hermaphrodites, with individuals bearing both male and female reproductive organs. Although it might seem that being simultaneously male and female could solve all your problems, in reality life for the average hermaphrodite is horribly complex. The problem arises because they still need to exchange sperm and eggs with other individuals to avoid inbreeding. But within each individual there is a conflict of interests between the male and female components. The hermaphrodite's dilemma is that the male component benefits by donating sperm to another because this increases the chances of fertilizing eggs. On the other hand, after being inseminated once, continuing to be a sperm recipient is costly because it results in tissue damage and the loss of control over fertilization. Nico Michiels and L. J. Newman[13] showed – in arranged encounters between lilac flatworms – that individuals go to extreme lengths to inject sperm and extreme lengths to avoid being injected. Individuals rear up, like miniature horses, and use the pointed penis projecting from the now vertical dorsal surface to stab and parry each other. The objective is to inject the other with sperm without itself being injected. There is no female sexual orifice; instead sperm are injected anywhere in the body surface since they can find their way to one or more of the numerous ovaries. In most encounters only one individual gets to inseminate and by doing so fertilizes eggs without incurring the cost of insemination wounds.

Our final example of traumatic insemination occurs in that almost mythical deep-sea creature, the giant squid. Instead of a hectocotylus (chapter 3) the male squid possess a large muscular penis. Giant squid are huge, weighing up to 220kg, and can have a combined body and arm length of 15m. At almost a metre in length, the penis is also relatively enormous, and through it the male injects a succession of elongated spermatophores under the female's skin. Since the spawning of giant squid has never been observed, quite how the sperm from the female's sub-dermal storage sites are released remains a mystery. One possibility is that she uses her suckers or her beak to cut herself and release the sperm. Few giant squid in reproductive condition of either sex have ever been examined, but a male caught off the Norwegian coast in the 1950s was found with several spermatophores embedded in the skin. Either he had shot himself in the foot during spawning, or more likely he had been riveted by another male during a spawning event.[14]

As I searched the literature for information about squid reproduction I came across an article in a Japanese medical journal describing the case of a middle-aged man who had gone to his doctor complaining of discomfort in his mouth and throat. Examination revealed dozens of tiny spindle-shaped structures embedded in the mucous lining of his mouth and pharynx, which the doctor thought were parasites. Detailed examination revealed their true identity – the spermatophores of a squid, which the man had eaten previously as sushi.[15] Several species of squid produce projectile spermatophores, which under normal circumstances are shot into the skin of the female; the one this man had eaten remains unknown. Mercifully, it was not the giant squid, whose 20cm-long spermatophores would have created considerably more than discomfort.

Copulatory Form

The process of coupling, mounting, mating, copulation, coition, bonking, whatever you want to call it, is almost infinitely varied across the animal kingdom. None the less most of us know what it is when we see it. The usual position in many species is for the male to be on the female's back. End-to-end copulation occurs in a few bugs, and squid and cuttlefish engage in head-to-head copulation. Face-to-face copulation is less common, but occurs in some crustaceans and an endangered New Zealand songbird, the hihi.[16] Apart from ourselves, bonobos are the only primates to copulate face to face. There is a certain irony to the fact that this position, once thought to be a defining human characteristic, is referred to as the missionary position, apparently because it needed to be taught to preliterate people.

Side-by-side copulation is more unusual still, but occurs in what must be the ugliest parrot in the world: the greater vasa parrot. During the two-day interview procedure for my present job I developed a rapport with one of the other candidates, Roger Wilkinson. He subsequently became curator of birds at Chester Zoo and almost twenty years later we made contact again when he wrote to tell me about the unusual copulation behaviour of the zoo's vasa parrots. All parrots, it seems, have fairly long-winded copulations, but the vasa is extreme, with the male and female forming a copulatory tie, just like dogs, lasting up to ninety minutes. During this time the birds sit side by side, connected only by the fleshy lobes of their cloacae. I was intrigued to see how this was achieved, so I wrote to various museums to see if any of them had a vasa parrot preserved in alcohol which I could borrow and examine. Fortuitously, there had been an expedition to Madagascar in the 1920s during which an ornithologist had shot a male vasa in breeding condition. He was so struck by the size of the male's cloaca that he cut it off and pickled it. The jar containing the dismembered protrusion then gathered dust in the Museum of Zoology at the University of Michigan until my search rediscovered it. The dissection was disappointing because the structure seemed to consist of little more than thin layers of tissue interspersed by sinuses and blood vessels. Later when I saw parrots *in copula* I realized that the protrusion was tumescent and could be engorged with blood at the appropriate time, thus accounting for its odd design. The vasa parrot is little bigger than a pigeon, yet at full size the male's cloacal protrusion is the size of a tennis ball. The way it works is this: the male's protrusion is inserted inside the female's cloaca and, as it swells, it locks the two birds together for the hour or so it takes to copulate. What is its purpose? Why do vasa parrots need such an elaborate, protracted copulation?

The answer, or at least part of an answer, came from a colleague, who on a trip to Madagascar had seen vasa parrots copulating in the forest. The vital point was that copulations involved *two* males. This coincided with the observations of parrot breeders who found that in captivity vasa parrots rarely if ever breed as a pair, but given additional males, would do so readily. Vasa parrots probably have a polyandrous mating system and if so, sperm competition is automatic, hence the need for behaviour and anatomy that maximize each male's chance of fertilizing eggs.[17]

Other species also show both very long and very short copulation durations. The most extreme example of protracted mounting occurs in a stick insect which remains *in copula* for seventy-nine days.[18] Obviously, it is unlikely that sperm transfer occurs throughout this prolonged union. Instead, remaining in genital contact is probably a form of mate guarding (see chapter 2). Copulation in other insects can last just seconds or minutes. Some fast copulations also occur in birds and mammals. Speedy sperm transfer in birds may be facilitated by the lack of a penis. None the less, the tenth-of-a-second copulation of the dunnock requires remarkable precision. Extremely rapid copulation is usual for a number of antelopes – presumably because those that took their time paid the price by ending up as a predator's lunch.

Rapid copulation can also be advantageous when competition and aggression between males for access to females is intense. When female primates copulate with subordinate males, they usually do so both furtively and extremely rapidly – to avoid the dominant male's wrath. However, the most extraordinary example of this occurs in a lizard: the marine iguana. When Darwin[19] first saw this species on the Galapagos he was scathing: 'It is a hideous-looking creature, of a dirty black colour, stupid, and sluggish in its movements.' Had he spent longer watching them he might have acknowledged that their extraordinary mating arrangement more than compensated for their unlovely appearance. On the black, wave-washed lava, dominant males defend tiny territories which together comprise a lek that females visit in order to copulate. Females copulate only once each breeding season so neither sperm competition nor sperm choice feature large here. However, competition for a female's single insemination is intense. Dominant males get most copulations but subordinate males have a curious strategy to increase their chances. They lie around the periphery of a dominant male's territory and attempt to waylay incoming females. The problem is that a normal iguana copulation takes at least three minutes. This is more than enough time for a dominant male to race over and wallop the subordinate.

To minimize the chance of this happening, subordinate iguanas ejaculate prematurely. Long before they have even got close to touching a receptive female, subordinate males partially ejaculate so that their viscous mass of semen lies ready and waiting at the entrance to their cloaca. If a female comes along they can inseminate her in much less than three minutes. This slick strategy isn't without its costs, however. A prematurely positioned ejaculate has a limited lifetime: once out of the safety of the male's epididymis, the sperm deteriorate rapidly, and if a male doesn't find a female he must, like Onan, cast his seed upon the larval rocks, and start again.[20]

Copulation Frequency

On another remote and windswept island – St Kilda in the Outer Hebrides – a field biologist has followed a female Soay sheep for five hours. During that time she has copulated no fewer than 163 times with seven different rams.[21] In the southern United States a virgin queen fire ant copulates just once in her entire life.[22] This isn't just a difference between sheep and ants; within both mammals and insects copulation frequency varies. Female chimpanzees are like Soay sheep in that they copulate many times each day when they are in oestrus, whereas female gorillas copulate only once or twice. Within insects, many bugs and flies copulate repeatedly, while others manage with a single insemination. It is a curious paradox that reproductive biologists, so obsessed with the mechanics of reproduction, have rarely considered why so much variation exists in the copulation frequency of different species. It has fallen to those with an evolutionary interest in reproduction to try to make sense of this variation.

There are, in fact, two rather different issues tied up with the question of copulation frequency: how often? and with how many different partners? Obviously if, like the fire ant, a female copulates only once, ever, there can be no sperm competition or sperm choice. The same is true even if a female copulates dozens of times, but always with the same partner. None the less, this variation still requires an explanation. But, as we have seen, both these scenarios are rare, and the females of most species copulate several times and often with different partners. What this means is that the questions of how often and with how many partners are inextricably linked.

A prerequisite for understanding the evolution of copulation frequency in its various forms has been accurate information on how often individuals copulate. It may seem surprising, but such information is difficult to obtain, and is available only for those species which can be easily observed, such as insects and birds. Our view of copulation frequency in mammals is distorted or, rather, obscured by the fact that many species are nocturnal. Information exists only for diurnal species that copulate in the open, like Soay sheep, some primates and seals.

We shall start with insects. In the late 1980s Mark Ridley, then at Oxford University, went through all the published accounts he could find regarding copulation frequency in insects.[23] Of 47 species that had been studied in detail, in at least 39 (83 per cent) females typically copulated more than once during their lifetime. Because of their lifestyle and the lack of any kind of pair bond, for most insects this means that females have multiple partners.

Birds on the other hand are like ourselves, with the sexes maintaining long-term social relationships which may be facilitated by repeated copulations. Most birds are socially monogamous and, even in those with a polygynous mating system, females often maintain a season-long relationship with just one male. This might appear to make them likely candidates for the many-copulations-but-with-one-partner prize, but as we saw in chapter 2 in most birds monogamy is an illusion: extra-pair copulations and extra-pair paternity are widespread. Since repeated copulation is one way in which a male can minimize the chances of being cuckolded (chapter 2), it should come as no surprise that some of the variation in copulation frequency between partners is closely tied up with the likelihood of sperm competition. At one extreme, some species, like the skylark, copulate just once or twice for each clutch of eggs but, at the other, goshawks and other birds of prey copulate 500–600 times for each clutch. Across birds as a whole this variation coincides with whether males perform mate guarding. In those species where males guard their fertile partner by close following, copulations between them tend to be infrequent. For those species where, for ecological reasons, males cannot guard, copulation frequency is much higher.[24]

The most convincing evidence that copulation frequency was closely linked to the likelihood of sperm competition in birds came from comparing species with monogamous and polyandrous mating systems. In the latter, where a female may have several male partners, as in the dunnock and the jacana, female copulation frequency is always very high. This is because, despite the apparent harmony between group members in these polyandrous systems, competition between males for fertilizations is extreme. In the dunnock the intensity of sperm competition is reflected in two extraordinary aspects of their behaviour and immortalized by the studies of Nicholas Davies of Cambridge University.[25] The first is a protracted precopulatory display in which the female stands still, shivering her wings, and raising her tail to expose a surprisingly red cloaca, reminiscent of the red 'bottom' of female primates. The male hops around in an agitated manner, pecking at the female's cloaca and in doing so stimulates the female to eject a drop of semen. Once the male has seen this, he copulates, flying at the female and touching her so rapidly you can hardly see it. The ejected semen is assumed to be that of a rival male, but this is yet to be demonstrated. The second adaptation to sperm competition is the rate of copulation. Female dunnocks often have two male partners, but they also breed regularly as monogamous pairs. The difference in the rate at which females in these different circumstances copulate is the most convincing evidence for a link between the frequency of copulation and sperm competition. A polyandrous female typically copulates over two hundred and fifty times with her two partners for each clutch, whereas females in monogamous pairs copulate only fifty times. Ignorant of all this, the Reverend Frederick Morris, writing in the 1850s, encouraged his parishioners to emulate what he referred to as the dunnock's 'modest lifestyle'. As a clergyman Morris was also vehemently anti-Darwinian, dismissing the *Origin* as the most 'inconclusive, illogical' book he had ever read.

The variation in copulation frequency among birds raises a number of questions. Initially it was assumed that males could dictate to females when copulations should take place. But it has since become clear that it is females and not males who control copulation. Males may initiate copulations, but females generally decide whether or not they will go ahead. The question then is this: if females are in control, and if they require only a few inseminations to ensure that their entire quota of eggs is fertilized, why do they comply with the male's demands and copulate so frequently? The only answer, and I am not completely persuaded by it, is that it is in the female's interest to convince her partner that he is the father of her offspring so that he will help to rear them. For some female birds total fidelity, imposed by repeated pair copulations, may be the price they have to pay to guarantee male assistance. Other females, however, may try to secure the best of all possible worlds – a caring partner plus extra-pair fertilizations. Females may achieve this difficult compromise in a simple but effective manner. They copulate frequently with their partner initially but then refuse all copulations from him once egg-laying starts. Since each egg is fertilized the day before it is laid, an insemination during the laying period is quite capable of fertilizing eggs. So by rejecting pair copulations at this time but accepting extra-pair copulations, should the right male come along, the female has considerable control over the paternity of her clutch.[26]

The social and mating systems of mammals differ from those of birds so it might not be a surprise to find that their copulation behaviours also differ. The main difference is that the males of most mammal species are polygynous and provide little or no parental care. This means that socially successful males will copulate with many different females during a breeding season, whereas females usually copulate only once or a few times. Female Northern elephant seals aggregate on the beaches of California and are monopolized by a dominant male – a harem-master who may be as much as seven times heavier than a female. A dominant male may inseminate thirty or more females during the ten-week-long breeding season, and over a hundred females during his two- or three-year tenure as top male. Most males, however, never get to copulate at all. Each female, on the other hand, usually copulates just once each year. She comes into oestrus and becomes sexually receptive about twenty-eight days after she has given birth, and is in oestrus for just two or three days. Fertilization follows soon after insemination but implantation of the fertilized egg is delayed for about three months so that the female gives birth eleven months later when she next returns to land.[27]

Biologists reckon that a mere 3 per cent of mammal species have a socially monogamous mating system (see chapter 2). The study of one of these, the prairie vole, by Sue Carter of the University of Maryland and her colleagues, has provided some tantalizing glimpses into the physiological correlates of monogamy. The prairie vole's mating system is rather like that of some birds and humans, comprising stable pair bonds which persist despite occasional extra-pair copulations by both sexes. Moreover, the bond between male and female prairie voles appears to be established through frequent copulation. In her first oestrus a female and her partner typically engage in bouts of copulation lasting up to forty hours. Since ovulation and fertilization occur about twelve hours after the start of oestrus, many of these copulations have nothing to do with fertilizing the female's eggs. In subsequent breeding cycles copulation bouts are less protracted, suggesting that the initial lengthy bouts serve to establish the bond. The fact that rodent species without monogamous bonds never engage in such protracted periods of copulation is also consistent with the notion that repeated intromission helps to establish the prairie vole's partnership. The clue to the way this works was provided by an understanding of the basis of motherly love. The bond between mother and offspring is facilitated by the hormone oxytocin which is released when a female gives birth or suckles her young and acts on the brain to generate maternal care. However, it turns out that copulation and close physical contact between partners also triggers oxytocin release. So the frequent copulation between recently acquainted prairie vole partners may result in a surge of oxytocin which helps to cement the close relationship between male and female.[28]

There are other mammals in which copulations appear to have an entirely social function. This appears to be true in many primates, including ourselves, but is taken to extremes by the bonobo. Also known as pygmy chimpanzees, bonobos are famous for their frequent promiscuous copulations and other forms of sexual activity, including male–male genital contact, and female–female genital contact. Bonobos of each sex typically copulate dozens of times each day. However, rather than employing copulation to cement a relationship with a particular individual, bonobos appear to use sex to facilitate social relationships with all their troop members. In bonobo society sex is a substitute for aggression. Social interactions that in other primate societies would normally lead to aggression are usually followed among bonobos by some form of sexual activity.[29] Why such a system should have evolved in bonobos and not in other primates remains a mystery. So far, no one has conducted a paternity study of bonobos, but the frequency of copulation is so extraordinarily high that sperm competition must occur – even though the majority of copulations appear to take place in a social rather than an overtly sexual context.

Sexual Conflict over Copulation

Since the number of copulations necessary to achieve fertilization differs for each sex, conflict over copulation is almost inevitable. Females require only one or a few well-timed copulations to ensure that their eggs are fertilized. But in species where females are polyandrous the likelihood of a male fertilizing a particular female's eggs depends on how many times he copulates with her. For many species, especially those with long-term pair bonds, male and female may reach some sort of compromise over how often copulations should occur, as in the birds of prey discussed earlier. But in others it is more obvious which sex has more control over copulation frequency.

A female butterfly uninterested in sex has only to twist her abdomen upwards to foil the most ardent suitor and a Soay ewe achieves the same result by holding her tail firmly downwards. However, the most accomplished and at the same time peculiar case of female control over copulation is provided by the spotted hyena.

Out on safari with your binoculars you would be hard pressed to tell a male hyena from a female. In the unlikely event that you were able to examine their genitalia in the hand, you would still find it difficult to distinguish male from female. The problem stems from the fact that the female's clitoris forms a fully functional mock penis. Ovid and other early naturalists were confused and bemused by this and thought hyenas were hermaphrodites. Another ancient and somewhat unscientific view was that the hyena's laugh reflected its mischievous delight at being able to change sex. However, they are not hermaphrodites; nor can they change sex. Instead, the clitoris is enlarged and erectile just like a penis; there is no vulva, the labia are fused together and the ureter passes directly through the clitoris. The arrangement is unusual, to say the least, since the female urinates, copulates and gives birth through this pseudo-penis. For many years evolutionary biologists have asked themselves how such an arrangement could have evolved. There are no firm answers, but there are some interesting and plausible ideas. For a mock penis to have evolved there must be some net benefit to females. The benefit must be considerable because females appear to pay dearly for their masculine looks, not least because in giving birth the clitoris is ripped apart as the young emerge. There are several possible benefits, however. First, females use their erect clitoris to signal their rank to other group members. Second, because the opening of the female's reproductive tract lies in an unusual position, at the tip of the clitoris pointing forward between her back legs, males find copulation extremely difficult. This gives females almost total control over whether a copulation takes place or not and, perhaps more importantly, whom she copulates with. A successful copulation depends on the acquiescence of the female and the nature of her relationship with particular males. As a result, males have little to gain from fighting with each other for a female's favours, and indeed aggression among male hyenas is rare. Instead males attempt to outdo each other in terms of being nice to females and not only when they are in oestrus. However, when a female is receptive, males perform protracted and pathetic fawning courtship rituals, cringing and whinging around the female in the hope that she will acquiesce and allow copulation to take place.[30]

The other extreme is where females have no control over copulation and are forcibly inseminated. For many people the film *Braveheart* (1995) was their first experience of *jus primae noctis* – the right of a medieval lord to copulate with someone else's bride on the night of her marriage. Judging from the web-page discussions which followed, medieval scholars were horrified by this. Not that it occurred, but by the fact that so much licence had been taken in making the film. Their feeling was that *jus primae noctis* was a myth – not least because any medieval lord could copulate with whoever he chose, whenever he liked, with little fear of retribution. In much the same way, Julius Caesar had 'legal' access to any wife in the Empire. But, however it occurred, *jus primae noctis* was rape. Any kind of rape is an obvious and horrific way in which sperm competition may occur. Several hypotheses have been proposed to explain rape.[31] One is that rape occurs merely for male sexual gratification. Another is that rape is simply violence against women. However, the fact that rape victims are significantly younger than female murder victims has been interpreted to mean that rape can sometimes have a reproductive function, although, of course, this is also consistent with the gratification idea. Discussing rape in an evolutionary context does not imply any moral or ethical approval, nor does it imply that all men are 'programmed' to rape: it merely reflects the possibility that it may be a way in which certain males achieve some reproductive success.

Rape is particularly frequent during war, a phenomenon for which there are at least two evolutionary interpretations. The first is that rapists are much less likely to be brought to account at such times. The second is that men who commit rape often have very short reproductive lifespans ahead of them. Even outside the context of war, rape remains disturbingly frequent, with over sixty thousand cases reported each year in the United States. Outside war those who commit rape tend to be low-status individuals who might not achieve any reproductive success by more conventional means. On the basis of the behaviour of certain aggressive preliterate societies, such as the Ache in South America,[32] it seems likely that in our primitive past rape may have been an important context for sperm competition.

Male coercion also occurs in a number of animal species. Forced copulation behaviour has been studied in detail in scorpion flies – so-called because their elongated abdomen curls up over their back and makes them look vaguely scorpion-like.[33] Males scavenge dead insects and other invertebrates, sometimes stealing them from spiders' webs, and present them to a female as a nuptial gift. If the female accepts the gift the male copulates with her as she eats it. The female benefits directly from the gift since the additional nutrients it provides enable her to lay more eggs. Males, however, sometimes struggle to find suitable gifts. In this situation they have two options. They can offer the female a gift of their saliva – not as good as a dead cricket, but better than nothing – or they can attempt to copulate forcibly with the female. It is perhaps not surprising that females are discriminating partners. They prefer males that offer the best deal and are distinctly unenthusiastic about copulating with coercive males and try to escape from them. However, males possess a structure, referred to as a dorsal clamp, on their back which helps them restrain an unresponsive female. Forced copulation is probably costly for females, not least because, without the nuptial gift, they produce fewer eggs. While males may appear to be the winners in this conflict, females do have some control. Females subject to forced copulation return to receptivity and are ready to copulate again much sooner than females who copulate with gift-bearing males and by doing so may be able to devalue the insemination of the coercive male.

Inhabiting one of the least hospitable habitats on earth, the Lake Eyre dragon has the least savoury sexual habits of almost any animal anywhere. Living in the hot, dry salt lakes of South Australia, males of this sexually dimorphic agamid lizard (males are up to 70mm in length; females 64mm) are able to bully unwilling females into copulating. Males rush at females, grasp them around the neck using their jaws and attempt to roll the female on to her side forcibly to inseminate her. An unreceptive female responds initially by trying to run away. If this fails and she is caught she arches her back and tries to prevent mounting. As a last resort she flips right over on to her back – a tactic that always thwarts males. It then becomes a game of persistence. If a male waits long enough the female will right herself and when she does so he will throw himself on top of her, preventing her from flipping over again and then inseminating her. Not surprisingly, the greater the relative size difference between the male and female the more successful males are at achieving forced copulation. However, forced copulation is potentially costly for females in two main ways. First, lying on their back in the open in an attempt to out-wait a male renders females vulnerable to predators. Second, some males are so aggressive that they kill females. In one incident observed by Mats Olsson a relatively large male waited for a female to right herself.[34] As she did so,

he threw himself on top of her, raped her while biting really roughly around her head and nape and then walked off. After he had left, the female lay still even as we approached closely, and we then realized that she was actually dead! The male's enlarged incisors had penetrated through the skull-base and killed her!

Surprisingly, forced copulations are not restricted to males imposing themselves on females. There are a few situations where females appear to browbeat males into copulating with them. The mantis shrimp, which inhabits the warm waters off the coast of Florida, is one such case. Its local name of 'thumb-snapper' comes from a unique style of foraging in which prey are stunned by the mantis shrimp snapping its forelimbs with great force – sufficiently great to be able to split open a human thumb. Female mantis shrimps are forward and aggressively harass males to copulate with them. In contrast males are shy, cautious and choosy – and with some reason, because they are sometimes injured by females during courtship. This is an odd situation, for a number of reasons. First, unlike some other species, such as phalaropes (see chapter 1) where females take the initiative in reproduction, female mantis shrimps are no bigger than males. Females also store sperm between successive spawnings, so it is unlikely that they are desperate for sperm to fertilize their eggs. It is therefore far from clear what females gain from inducing males to copulate with them. The only plausible benefit is that females obtain nutrients through male ejaculates, but this remains to be tested.[35]

Fertilization

Success! Fertilization is equivalent to the moment you learn that you have the winning lottery ticket. Well, almost. From my point of view, success in a lottery draw would be having the cash in the bank or, better, in used notes, rather than simply being told I had won. Similarly for animals, reproductive success is the production of a viable offspring, not just fertilization. However, fertilization is an important first step to success.

From the moment a female first decided to copulate, all her behaviour and physiology have been geared towards making sure that not too many and not too few sperm make contact with her ova. This regulation continues right up to the point of fertilization: for most organisms the penetration of an ovum by more than one sperm (a phenomenon known as polyspermy) means embryo death. In mammals the ovum leaves the ovary surrounded by a cloud of cells, the cumulus, through which sperm must pass. This is another trap: the cumulus contains traces of fluid from the ovary which causes some sperm to react prematurely. The follicular fluid triggers the acrosome reaction, rendering these sperm incapable of fertilization. So near and yet so far. Any sperm that survive this treatment still have to get past another inner layer of cells, the corona radiata, and then beat their way through to the covering of the ovum, the zona pellucida. The sperm head and the zona fuse together, and the sperm's hyperactivated flagellum drives the head of the sperm through the zona. The tip of the sperm then explodes (the acrosome reaction), exposing the receptors on the head of the sperm. These enable the sperm head to fix itself to the inner layer of the ovum. As the contents of the sperm spill out and mix with those of the ovum the sperm's flagellum ceases to beat; it breaks away and is eventually left outside the ovum. It is at this point that the processes that prevent polyspermy occur: chemical messengers released from within the egg prevent further sperm from gaining ingress. Embryo development is now given a kick start as the sperm triggers off waves of calcium release within the egg. At the same time, the sperm's tightly packed chromosomes unravel to become the sperm pronucleus and the same thing happens to the ovum's chromosomes. Some twenty-four hours after the sperm first encountered the egg the male and female pronuclei can be seen lying side by inside the ovum. Although this is the point many reproductive biologists consider to be fertilization, the pronuclei still have to fuse and begin the first cell division to become a zygote.[36]

It has taken years of research to glean the most basic understanding of the complex process of fertilization and there are still many aspects to be explained. One of these may have a special relevance in the light of sperm competition and sperm choice. All the cells in our bodies are said to be diploid because they contain a double set of chromosomes. Our gametes are the only exception. Produced by a process of cell division, referred to as meiosis, they have only a single set of chromosomes, and are said to be haploid. The logic is simple: at fertilization the haploid sets of male and female chromosomes combine to give the new zygote a full diploid complement. Sperm attain their haploid status, days or weeks earlier, while they are being formed in the testes. One might assume that the same would be true of ova in the ovary but this is not the case. Ova are diploid when they leave the ovary, and indeed retain their diploid state almost up to the moment before fertilization. The mystery is why the ovum retains both sets of chromosomes for so long. A fascinating suggestion, by Claus Wedekind[37] of the University of Bern, Switzerland, is that this has evolved to provide an additional opportunity for the female, or rather her ovum, to modify the genetic quality of the zygote. After the sperm has penetrated the outer layers of the ovum and its chromosomes have unravelled to form the male pronucleus, the ovum has the opportunity, Wedekind suggests, to assess the sperm's genetic make-up and 'decide' with which of her two sets of chromosomes it is most compatible. The ovum then undergoes its final division, jettisoning the less appropriate chromosome set, allowing the remaining set to combine with those of the male.

There is one further process essential for successful embryo development – the destruction within the fertilized egg of the sperm's mitochondria. It is possible that even during the sperm's brief life its mitochondria may have been damaged by metabolic processes and their incorporation into the developing embryo could be fatal. Even after two sets of chromosomes have fused and formed a new zygote, a female still has the opportunity to modify her choice of father for her offspring. Whether this zygote ever becomes an embryo or, eventually, a reproducing adult, depends on the female. A female has the potential to reject a fertilized egg or she may allocate more or less nutrients to it during its development – depending on its genetic composition, among other things. Other than knowing that genetically incompatible zygotes are often aborted,[38] this is an area we still know little about and it remains to be seen how extensive this form of post-fertilization female choice is.

In the last three chapters I have described the machinery associated with reproduction and the processes by which gametes are produced and by which a few of them eventually meet. Reproductive physiologists and anatomists, particularly those studying mammals and birds, accumulated all of this information with the implicit assumption that the sperm in a female's reproductive tract came from only a single male. As we have seen, in most cases this is unlikely to be true, so in the following chapter we look at what it is that determines which of several males fertilizes a female's eggs.

6 Mechanisms of Sperm Competition and Sperm Choice

And cavaliers and noblemen came before her, and one of them, whichever she might choose, had to lie with her before the bridegroom, and if she became pregnant from that knight the son she bore would be noble, otherwise the children of the husband were commoners.

R. BRIFFAULT, *The Mothers* (1927)

When Bateman conducted his fruitfly experiments in the 1940s (chapter 1)[1] he used different mutants to establish genetic markers so he could see which male fertilized a female's eggs. If Bateman had recorded the order in which different males copulated with each female he would have found that in almost every case the last male to inseminate a female fathered most of her offspring. How does this last-male sperm precedence arise? What are the underlying mechanisms that result in the last male being so successful? There are two main possibilities. It could be due to some sort of contest between the sperm from the two males; that is, classical Darwinian male–male competition, but taking place out of sight inside the female reproductive tract. Or it could be due to female choice; that is, the female prefers the last male, or his sperm, and she therefore manipulates the ejaculates inside her in such a way that his sperm are used preferentially. A further possibility is that some combination of these two processes is responsible for the outcome.

Trying to sort out just why it is that the last male fertilizes most eggs has been very tricky and research in this area is still going on. Insect biologists had a number of ideas about possible mechanisms that would produce last-male sperm precedence. However, to see whether their ideas were sensible they had to know something about the way the female reproductive tract was designed and the way sperm were utilized by females.

Luckily, there has been a long history of anatomical description in entomology and so, for many insects and other invertebrates, the basic structure of the reproductive system was often already known. It was also apparent from the studies of early insect anatomists that the variety in the design of both female and male reproductive equipment was mind-boggling. This is hardly surprising: there are after all some 750,000 species of insects. But this diversity in structure meant that it was unlikely that there would be universal rules governing sperm utilization across all species. None the less, one general observation was that the females of most species possess a sperm store, usually lying just off to one side of the egg-duct, enabling females to release sperm on to the eggs as they passed down the oviduct. Understanding the way females stored and released sperm might provide the clue to last-male sperm precedence.[2]

One of the simplest mechanisms is 'last in, first out'. The sperm from the last male to copulate push those from previous males to the back of the female's sperm store, or spermatheca. By doing so they ensure that they are nearest the exit and ready to be first out when eggs need to be fertilized. This idea has an intuitive appeal, but demonstrating that this is what actually happens is extraordinarily difficult. In a few species this mechanism seems very likely, albeit with an additional piece of trickery. Ghost spider crab females copulate with several males and the last male almost always fertilizes the eggs. The way this is achieved is a testament to the evolutionary pressures on males created by sperm competition. When the male ghost spider crab copulates with a female he first transfers seminal fluid and no sperm. This seminal fluid is effectively epoxy resin: it goes into the female's sperm store, pushing any previously stored sperm to the back, and then sets hard. The resin seals in the other males' sperm. Only then does the male introduce his own sperm, and providing he is last to inseminate the female before she fertilizes and lays her eggs, he will father them all.[3] Without the epoxy resin trick, the sperm from successive ejaculates would mix, and the last male would fertilize only a proportion of the eggs. The giant water bug, which we met in chapter 2 (p. 55), achieves last-male precedence by using the 'last in, first out' system, but without the glue.[4]

Another way that the last male could fertilize most eggs is by removing the sperm from rival males from within the female. This sounds implausible, but in one group of insects it is *de rigueur*. In damselflies and dragonflies over 90 per cent of offspring are attributable to the last male to copulate with the female. The penis of these beautiful insects is adorned with horns and little hooks which they use to drag from the female the sperm of previous males. Copulation is brief, lasting only ninety seconds, and during this time the male can be seen making pumping movements. Jonathan Waage separated male and female damselflies at different stages of copulation to work out what was going on. Even before copulation started the female's sperm store was full from previous inseminations. After copulation had proceeded for thirty to forty seconds he found the sperm store empty, but by the time copulation was complete it was full again. Waage wondered whether the male was removing stored sperm prior to transferring his own. He confirmed that this is what was happening when he saw masses of sperm adhering to the hooks on the penis, following the disappearance of sperm from the female's store.[5]

This and subsequent studies of dragonflies seem to epitomize the view, prevalent during the 1970s among those studying sperm competition, that females were essentially passive receptacles for male sperm. Certainly it seemed from the damselfly study that females had little influence over copulation or its outcome. More recently, as we shall see, the role of the female in determining the outcome of sperm competition has been taken more seriously.[6] The initial emphasis on male perspectives probably reflected the fact that most of the individuals studying sperm competition at that time were male. However, politically inappropriate though that may now seem, I believe it was fortuitous. If we had known right from the outset that both male and female factors could affect the outcome of sperm competition, resolving their relative importance might have seemed overwhelmingly complicated. As it is, because we dealt first with male factors, and only later started to consider what role females might play, things were kept sufficiently simple for resolving the problems to appear tractable.

A truly extraordinary mechanism of last-male sperm precedence occurs in a rather ordinary-looking beetle, studied by a German biologist, Klaus Peschke.[7] Rove beetles perform an end-to-end copulation and remain together for about forty minutes. Half an hour into copulation the male transfers a rather special sperm package or spermatophore into the lower part of the female's reproductive tract. Shortly after copulation is terminated and the male and female separate, the spermatophore takes over – with what appears to be almost a life of its own. A tube extrudes from the spermatophore, passes up the female tract and into her sperm store (spermatheca). Once there, the tip of the tube blows up like a balloon and almost completely fills the sperm store. As it expands, the balloon pushes the sperm from any previous inseminations back down the spermathecal duct and out of the way. Meanwhile, lurking inside the balloon are the male's own sperm, and these are set free into the female's sperm store by the female herself. On the inside of her spermatheca are two 'teeth' which she uses to puncture the balloon and release the sperm. Peschke's results are all the more impressive because through the use of sophisticated microscopy he has been able to see (and video) all this happening directly.

The processes that result in last-male sperm precedence in the ghost crab, in dragonflies and damselflies and in the rove beetle have been elucidated in part because it was possible to devise ways to see what was happening. In other organisms this has not been possible, and investigators have had to use other techniques. The dungfly provides a good example of this alternative approach. Parker knew from his sterile male experiments that the last male fertilized about 80 per cent of a female's eggs.[8] Because it was impossible to follow the sperm from individual males inside the female reproductive tract, Parker decided to tackle the mechanism issue in another way. Thinking like a plumbing engineer, he devised processes in which two fluids could flow in and out of a system and result in last-male precedence. He built mathematical models of plausible sperm ebb and flow processes and then conducted experiments with the flies to determine which model best agreed with the pattern of paternity he saw. The simplest model, which Parker referred to as a 'fair raffle', assumed that the sperm from two males went into the female's sperm store and mixed at random. Each sperm was equivalent to a raffle or lottery ticket, and each male's chance of winning, or fertilizing eggs, was proportional to the number of sperm he inseminated. It was immediately obvious that this was not the situation for dungflies, because, as we have seen, one male usually fertilized most eggs. A second model, the 'loaded raffle', assumed that the tickets of one male had some sort of advantage. One possible advantage might be that some of the other males' tickets have gone out of date by the time the result of the raffle is decided. This seems reasonable; sperm have a finite life inside a female's store, so it is more than likely that some sperm from an earlier insemination will have died or become less effective with time, giving the most recent male an advantage. However, this was not consistent with the last-male precedence pattern in dungflies because even with an interval of just a few minutes between copulations the last male still fertilized most eggs.

The third model was one in which incoming sperm shunted previously stored sperm out of the female's store but after a while started to mix with stored sperm. Imagine a swimming pool with an in-flow and an out-flow. Hot water pumped in at one end will push the cooler water out at the other, at least initially. After a while, some of the hot-water molecules will become mixed up with the cold ones, and will also be pushed out by the incoming warm water. By the time most of the water in the swimming pool is hot, most of the out-flow will also be hot. In quantitative terms, the relative amount of hot water in the pool will increase with exponentially diminishing returns. Now imagine the swimming pool is the female's sperm store and that there is only a single connecting tube which serves as both the in-flow and out-flow. Male 1 fills the swimming pool. As male 2 inseminates the female, pumping sperm into her spermatheca, the sperm of previous males is pushed out. As male 2 continues to copulate, displacing most of the previous sperm, he eventually starts to push out some of his own sperm. In other words, beyond a certain point it becomes increasingly difficult to remove the last few sperm inseminated by previous males. Although this mental model of the in-flow and out-flow of sperm from the dungfly sperm store is consistent with the observed pattern of sperm precedence, it seemed biologically implausible, to me at least. I simply could not envision two fluids in the same tube flowing in opposite directions.

Recently, two different approaches by two research groups found what appears to be the true mechanism that results in last-male sperm precedence in the dungfly. David Hosken and Paul Ward at Zurich University pieced the story together from a detailed study of the female reproductive tract at different stages during copulation. Leigh Simmons at Perth University and Geoff Parker, on the other hand, used radio-tracers to follow the fate of each male's sperm inside the female tract. The fact that both teams reached similar conclusions using these different methods gives me considerable confidence that they are correct. Starting with a virgin female, at insemination the male deposits his semen in the female's bursa – a bag-like structure connected to the sperm store by a long narrow duct. Soon afterwards a piston-like arrangement sucks up the sperm into the sperm store. So far so good – the female's store now contains sperm. The female now copulates with another male. The internal process repeats itself, but now the piston blows most of the original ejaculate out of the sperm store and back down the duct into the bursa before sucking up the semen from the second male. An important component of the second male's success is how well the tip of his penis fits over the entrance to the duct between the sperm store and the bursa. A good fit ensures that the previous male's sperm is blown out before the second male's sperm is sucked up into the sperm store.[9] This is a Mickey Mouse account of the sperm competition mechanism in the dungfly, inasmuch as it oversimplifies the processes. As we shall see later, the two research groups are in fierce disagreement about the details and, in particular, the extent to which females influence the uptake and utilization of sperm.

Nobody Here but Us Chickens

Among birds, just as with insects and some other invertebrates, the last male to copulate with a female fertilizes most of her eggs. Remarkably, this was known even in Aristotle's time. He wrote: 'The bird who carries an egg conceived by copulation, if it then has coitus with another male, will hatch all its eggs of a genus similar to the latter male.'[10] Aristotle's observation simultaneously encapsulates both the existence of sperm competition in birds and its outcome: last-male sperm precedence. However, the biological significance of these observations remained obscure until about fifty years ago when poultry biologists, motivated by the need to improve output, started to unravel the complexities of chicken reproduction. Two studies provide the starting point for the understanding of sperm competition. The first, by Paul Martin and colleagues at the University of Illinois,[11] revealed that if a female was artificially inseminated once with a mixture of sperm from two males mixed in different proportions, the ratio of offspring fathered by the two males was almost exactly that predicted from the number of sperm from each male. The second study, conducted by Mark Compton and his colleagues at Virginia State University,[12] showed that when females were inseminated twice, four hours apart, with equal numbers of sperm from two breeds of fowl, the second male fertilized most (about 77 per cent) of the offspring. Taking these two studies together it looked as though the factor that determined whether there was a last-male effect or not was the time difference between two lots of sperm entering the female tract. So if two males contributed equal numbers of sperm in a single mixed insemination, or in two inseminations made immediately one after the other, paternity was shared equally. But with a four-hour gap between equal-sized inseminations, as in Compton's experiment, the second insemination always did much better. By way of an explanation it was suggested that if inseminations were close together the sperm from the two males mixed before they went into the female's sperm storage tubules, but if the inseminations were separated by a few hours, the sperm from the two males remained stratified within the female's storage tubules.[13] 'Last in, first out' again. The hen's sperm storage tubules looked perfect for such an arrangement – they were long and thin, and it was easy to imagine sperm from successive ejaculates stacked up inside them.

There was a major problem with this interpretation, however. If stratification was indeed the cause of last-male precedence, then the chicks hatching from eggs laid earliest in the clutch should have been fathered by the last male to copulate, and the later-hatching chicks should be fathered by the first male. But no such pattern was apparent.[14] The true mechanism was revealed by looking at the way hens use the sperm given to them by males. This was worked out by a Scottish poultry biologist, Graham Wishart, using an ingenious technique.[15] Recall that fertilization takes place at the top of the oviduct, in the infundibulum, where sperm wait for the chance to penetrate the ovum (see chapter 3). At the moment the ovum is released from the ovary (see figure 2 above, p. 61), it is covered by a thin skin called the outer perivitelline layer. Fertilization occurs within a few minutes as sperm pierce this skin and one of them fuses with the female nucleus within the ovum. Immediately afterwards the cells of the infundibulum start to secrete a second skin, the outer perivitelline layer, around the ovum. As this second layer is laid down all the sperm remaining in the infundibulum are trapped between the two perivitelline layers. Wishart devised a means by which these sperm could be seen and counted. It involved taking a freshly laid egg, removing the layers that surround the yolk and examining them under the microscope. The trick was to use a fluorescent dye, which made the DNA in the sperm nuclei light up like electric blue cigars. That's the technical bit. The other part of the process was to obtain a series of eggs from a hen that had been inseminated just once. As sperm are released from the storage tubules and travel up to the infundibulum to fertilize each egg in turn, the numbers associated with successive eggs declines. The key point is that the *rate* at which the sperm are released from the tubules is constant. This means that initially, when there are a lot of sperm present, large numbers appear on the perivitelline layers of the first egg, but as the sperm are used up each successive egg contains fewer and fewer sperm (figure 11). The decline in the

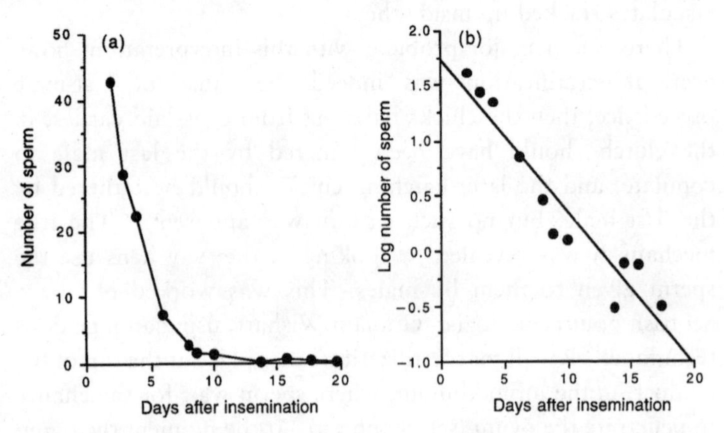

Figure 11

The decline in the number of sperm on successive eggs in the domestic fowl following a single insemination. The figure on the left shows the absolute numbers of sperm on a fixed area of yolk tissue, and the figure on the right shows the same information expressed as the logarithm of sperm numbers. The slope of this relationship describes the rate of passive sperm loss from the female reproductive tract (from Wishart (1987)).

sperm numbers on successive eggs provides a measure of the rate at which sperm are lost from the storage tubules.

This pattern of sperm utilization provides a rather simple explanation for last-male sperm precedence. Imagine a female inseminated on two separate occasions with the same number of sperm. Soon after the first insemination has occurred, sperm begin to leak from the storage tubules. By the time the second insemination has taken place most of the sperm from the first insemination have been lost and are outnumbered by those from the second (figure 12). This process is referred to as passive sperm loss, and I predicted that if this accounts for last-male

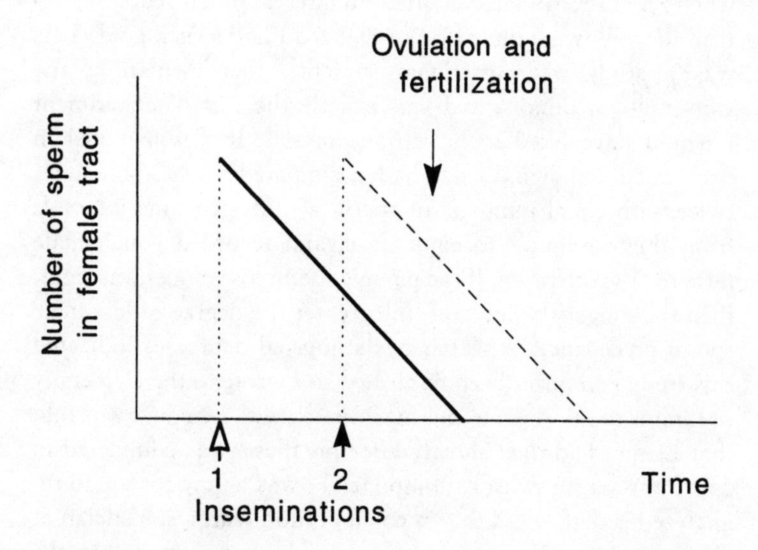

Figure 12

A graphical representation of the passive sperm loss model which results in last-male sperm precedence in birds (from Birkhead, 'Sperm competition in birds' (1998)). Two inseminations containing equal numbers of sperm are made (indicated by arrows 1 and 2) some time apart. Once inseminated, sperm start to be lost from the reproductive tract at a constant rate. By the time ovulation and fertilization occur (downward arrow), most of the first male's sperm have been lost from the tract, and so the second male's sperm are numerically dominant and fertilize most eggs.

precedence, then the longer the interval between two otherwise identical inseminations, the more marked the last-male effect should be – simply because the more time between the inseminations, the more time there has been for sperm from the first insemination to be lost.

All I needed was a way to test this idea. I toyed with the idea of setting up my own chicken colony, but a fortuitous event saved me the trouble. I was rummaging around in a box of papers on sperm competition when I found a photocopy of a Ph.D. thesis that I had been sent ten years previously. I remembered how I hadn't been able to make head or tail of it when I first received it. But, after an interval of ten years, as I re-read the yellowing photocopy I realized I had struck gold. This was a study by a graduate student, Allen Leman at the University of Illinois, and was exactly the sort of experiment I would have liked to have done myself. It was a model in experimental design. Leman had inseminated batches of females twice with equal numbers of sperm at different time intervals from thirty minutes to eight days, and recorded which male fathered the offspring. If the passive sperm loss model was right, then the longer the interval, the greater the degree of last-male sperm precedence. A search of the journal databases indicated this study had never been published so I wrote to the University of Illinois to ask them to put me in touch with Leman. I was told that Leman had died shortly after his thesis was completed in 1975, but his supervisor, Philip Dziuk, was happy for me to re-analyze his data. I did this in collaboration with a statistician at Sheffield, John Biggins, and our analyses confirmed that the passive sperm loss model was the basic mechanism: the match between what our model predicted and what Leman had found was reassuringly convincing.[16]

A Freeze-dried Female

There are two major categories of bird: passerines – also known as songbirds – which includes species such as blackbirds and robins; and non-passerines, which include the chicken and its wild ancestor the jungle fowl. Although sperm competition almost certainly occurs in wild jungle fowl, in most non-passerine birds polyandry and sperm competition are rather infrequent (see chapter 2). It is among the songbirds where sperm competition is most intense. I felt it was important, therefore, to see whether the mechanisms that resulted in last-male precedence in chickens also occurred in a passerine bird.

To pursue this I needed a species in which extra-pair paternity occurred in the wild and one I could breed easily in captivity. As a boy I'd kept zebra finches, and I knew that these tiny, prolific Australian birds would be perfect. Zebra finches come in a number of colour forms so I also realized that it would be possible to use plumage markers to determine paternity. In addition, extra-pair copulations were known to be frequent in the wild, and some of them resulted in extra-pair paternity. Our experiments with captive birds revealed that, just as with domestic fowl, the last male to inseminate a female fertilized the majority of eggs. In terms of getting to grips with the mechanism, however, there was a problem. In our studies of chickens we had used artificial insemination to control the timing of inseminations and the number of sperm. With the zebra finch we had to rely on natural copulations, and it was difficult to establish how many sperm males transferred during copulation. This was something, it transpired, that was not known for any species of bird.

We found out almost by chance. I remembered reading about a male sparrow that had been seen attempting to copulate with his partner who had just been killed by a car and I wondered if we could persuade a male zebra finch to copulate with a dead female. In a casual way, I posed this question to my assistant Jayne Pellatt. Unbeknown to me she went off and, over a matter of months, developed and perfected the technique. One day she came into my office and announced that she'd cracked it. She told me what she'd done and offered a demonstration. A freeze-dried female zebra finch fitted with a false cloaca was fixed on to a perch and placed inside the cage of a male zebra finch. With unabashed ardour the male went into his courtship routine, mounted the female and left his tiny ejaculate inside the false cloaca.[17]

Exploiting this new technique, we figured out what factors affected a male's ejaculate size. There was substantial variation in the numbers of sperm transferred, from less than 1 million to over 10 million, and some of this variation was down to how often the male ejaculated. If we used a male who had not copulated for a few days, he ejaculated many more sperm than one that we had used the day before – which wasn't really surprising. In addition, we established that the rate at which sperm were lost from the female zebra finch's sperm storage tubules was higher than that for chickens. And we made video recordings of the birds to see how often they copulated. Combining this information with our newly acquired knowledge of sperm numbers, we found that the passive sperm loss model satisfactorily accounted for last-male sperm precedence in the zebra finch.[18]

In conclusion, the basic mechanism of sperm competition appears to be similar in both the zebra finch and the domestic fowl, and hence most passerine and non-passerine birds. Our experiments show that the outcome of sperm competition depends on a number of factors in addition to the rate at which sperm are lost from the female reproductive tract. These include the number of sperm each male inseminates, the interval between inseminations and, as we found from other experiments, the timing of insemination relative to egg-laying. As well as these male attributes, it is possible that there are as yet undiscovered female attributes that influence paternity.

Mammals

I once owned a beautiful female Siamese cat. As with many breeds generated by intense selective breeding, mine was typically slow-witted. Her lack of behavioural finesse was especially apparent when she was on heat, since I regularly witnessed her in the garden surrounded by a motley assortment of monstrous males, each of whom took his turn to copulate with her. As I watched I wondered whether, as in insects and birds, some order effect would determine which of these males was the most likely to father the kittens. For a tom-cat seeking to maximize his production of offspring is it best to be first or last in the copulation queue?

For a long time the answer to this question depended very much on whom you asked. Although there are no studies specifically dealing with sperm competition in cats, studies of rabbits and guinea pigs showed that it was best to be first. On the other hand, in prairie voles the second of two males fathered more offspring. Yet other studies, such as those involving rats, found no advantage associated with being first or last. These results indicated that there might be no general mechanism for mammals and that each species has its own rule.[19] Obviously, in those mammals where females ovulate in response to copulation, so-called induced ovulators, we would expect the first male to father the majority of offspring. And this is precisely what happens. Thirteen-lined ground squirrel females are receptive for an incredibly brief time, just one afternoon each year. Males tear around looking for unmated females, hoping to be first in the queue. A male that finds a female already accompanied by another male has a dilemma: does he join the queue and hope to copulate second and have at least a small chance of fathering offspring, or does he continue to search in the hope that he'll find a lone, unmated female? The answer is that males often queue.[20]

Among females of species where ovulation is spontaneous – that is, not induced by copulation – things are rather different. The ground rules were established through a study of a household pet, the hamster. In the wild, hamsters live in Eastern Europe and Russia; for most of the year they lead solitary lives in underground burrows. The home range of a female may overlap those of several males, so in the wild the chances are that females have the opportunity to copulate with more than one male. When females come into oestrus they are sexually receptive for about twelve hours before and another nine hours after ovulation. In captivity they will copulate with several males during a single period of oestrus. William Huck and colleagues at the Sangamon State University, Illinois,[21] performed a cleverly designed experiment to establish which factors determine the fertilization success of two males copulating with a single female. The paternity of offspring was established using coat colour, albino and beige, as a genetic marker. What they found was that when the female copulated with two males relatively early in her oestrus cycle the second male consistently fathered more offspring, regardless of the interval between the two inseminations. On the other hand, when the first insemination took place at around the time of ovulation the first male fathered most offspring, and this effect was greater the longer the interval between inseminations. In other words, whether there was a first- or second-male advantage depended on when the inseminations occurred relative to ovulation. A male's best tactic is to inseminate a female at a time that ensures that his sperm are ready and waiting at the top of the oviduct just when the female ovulates. Let's assume sperm can get from the point of insemination to where fertilization occurs in five minutes. Naïvely, we might suppose that a male's best strategy is then to inseminate a female five minutes before she ovulates. There are two problems with this. First, neither male nor female knows exactly when ovulation will occur, so it is extremely tricky to time copulations precisely. Second, and perhaps more importantly, the male's sperm take time to get themselves ready for fertilization. The interval between ejaculation and capacitation varies markedly between species, from ninety minutes in mice to about three hours in humans. This means that to have the best chance of fertilizing a female's eggs in a sperm competition situation a male must time his insemination so that his sperm have just capacitated by the time the female ovulates. Golden hamster sperm take about three hours to capacitate, after which they remain viable for a further thirteen hours. Hamster ova also have a finite time period over which they can be fertilized – six hours after ovulation only half are still viable. Synchronizing the appearance of capacitated sperm with that of fertilizable ova is the secret of success.

To put their mechanistic information on sperm competition into perspective, William Huck and his collaborators also looked at the behaviour of male and female hamsters. They created a near-natural set-up in a large enclosure which allowed them to watch what happened in the underground burrows. The night the female's oestrus started the dominant male remained inside her burrow, sleeping between the female and the entrance, thereby giving himself the opportunity to repel other males. Interestingly, even though the female was receptive as soon as her oestrus started, the male was restrained and did not begin to copulate until later, about six hours before she ovulated. The male continued to copulate with the female on and off for one or two hours, even after his sperm supplies were exhausted. Thereafter he followed the female for a further two hours. This combination of behaviours would have minimized the chances of the female going off and copulating with another male. Even the spermless copulations may have served a purpose, by reducing the female's interest in further copulations.

The hamster provides a general model for how sperm competition is thought to work in mammals with spontaneous ovulation.[22] The basic idea is shown in figure 13 and can be summarized by saying that when two males copulate with a female their chances of fathering offspring depend on the interaction between the order of copulation, the interval between the two copulations and the timing of the inseminations relative to when the female ovulates.

The sperm competition processes I have just described for one insect, birds and mammals are all the *basic* mechanisms. Overlaid on top of these are other, much more subtle, processes. We can categorize them crudely into two groups: male and female processes. The main male subtlety is the difference

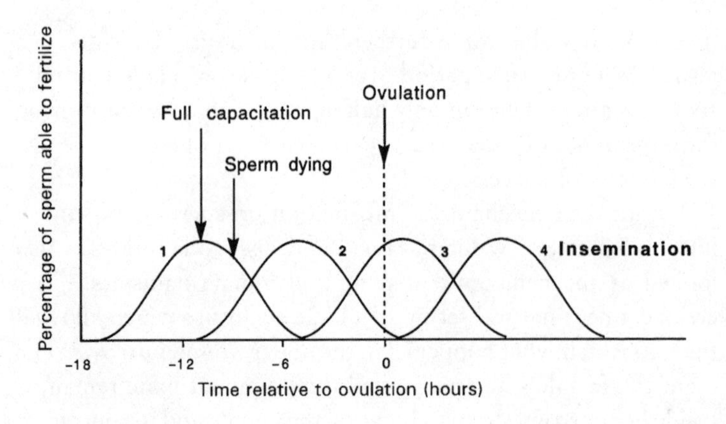

Figure 13

A graphical representation of the mechanism of sperm competition in mammals (from Ginsberg and Huck (1989)). Four hypothetical inseminations (1–4) are shown: for each one the percentage of sperm capable of fertilization is shown by a normal distribution, increasing initially as more sperm capacitate and then declining as sperm die. Sperm in the first two inseminations have peaked and died before the female has ovulated (time 0). The sperm in the fourth insemination were inseminated at the point of ovulation and therefore peak long after ovulation. The third insemination takes place six hours before ovulation, but as a consequence the sperm reach their peak close to the time of ovulation and hence are most likely to result in fertilization.

between individuals in their ability to fertilize eggs. The female subtlety is their ability to discriminate between, and differentially utilize, the sperm of different males. We shall start by discussing the differential fertilizing capacity of males, and then go on to consider in rather more detail the controversial subject of cryptic female choice.

Differential Fertilizing Capacity

Animal breeders are quite intolerant of any individual that doesn't come up to scratch in the reproduction stakes. Culling of males is more ruthless than that of females because a single good male can inseminate dozens of females. Breeders regularly screen their males: selecting the best and rejecting the rest. Screening males for their fertilizing ability can be a slow process – scoring conception rates, the proportion of females giving birth and litter size. Then, in the 1960s, it was recognized that a more rigorous form of screening could be achieved by mixing the sperm of different males in equal numbers, artificially inseminating it and looking at which male fathered most offspring. The earliest study of this kind was conducted as long ago as 1914 between different strains of rats and revealed that normally coloured rats had a fertilization success eight times greater than their albino competitors.[23] In some cases the differences between strains or individual males were even more dramatic. One study of turkeys involved mixing the sperm of fifteen different toms in equal proportions and inseminating dozens of females: only one of the males fertilized any of the eggs! Similar, if somewhat less extreme, results from the same type of experiment have been obtained with almost all domesticated mammals, including pigs, cattle, sheep and rabbits, enabling investigators to rank males in terms of their fertilizing ability. From a practical perspective the rank order achieved by this method was the same as that derived from single insemination studies – that is, with no sperm competition – but experiments in which sperm were mixed, so-called heterospermic inseminations, were over one hundred times more sensitive at detecting differences between males. This effect is referred to as differential fertilizing capacity. One of the people who played a central role in conducting these studies was Philip Dziuk at the University of Illinois.[24] He drew an analogy between ranking the fertilization success of males and the scoring success of basketball teams. The number of 'baskets' scored by each team during the warm-up period when a team is by itself might provide a reasonable picture of their ability, but a much more revealing test is to look at the scores when two teams play each other. Differential fertilizing capacity was obvious in the sperm competition experiments we undertook on chickens and was something we had to take into account when calculating the fertilization success of each male.

The most notable aspect of differential fertilizing capacity is that, despite being so well established, it is poorly understood. Animal breeders have carried out dozens of experiments that unequivocally demonstrate its occurrence. Usually, the males that rank highly in the heterospermic stakes have better-quality sperm, faster moving and with fewer deformities – giving them, as we saw in chapter 4, a considerable edge in a competition with those of other males. But what is remarkable is that differential fertilizing capacity still persists despite generations of intense artificial selection by animal breeders for superior fertilizers. With such strong directional selection for fertilization success, all the variation in differential fertilizing capacity should be used up, but it isn't. The processes that maintain variation in traits like this remain a mystery.

What animal breeders have *not* done is any experiment to see whether females show a preference for the sperm of particular males. To establish this, breeders would need to have inseminated the same individual females over several different breeding cycles with the same mixture of sperm to see whether females showed any consistent differences between one another in the paternity of their offspring. The reason animal breeders never did this was because it is irrelevant to the mass production of chickens, pigs or cattle. But from an evolutionary biologist's perspective, it is crucial to know the extent to which females influence the fertilization process.

Whose Sperm Is It Anyway? Cryptic Female Choice

Terry Roberts is a builder whose hobby is breeding birds. Not just any old birds. His aim is to breed a hybrid between a chaffinch and a canary. In the nineteenth century coal miners used canaries to warn them of poisonous gas underground: if the canary fell off its perch it was time to get out. Miners became fond of their birds and as well as breeding them with each other, they started to produce crosses, or mules as they called them, with British finches such as goldfinches and greenfinches. Finch mules have always been more difficult to breed than canaries or finches themselves, but some were easier than others.[25] The ease with which a mule can be produced depends on the evolutionary similarity between the two species. The canary is, after all, a kind of finch. The greater the genetic similarity between the two species the easier the mule is to produce. The reason why the chaffinch–canary mule is so elusive is that the chaffinch is genetically more dissimilar to the canary than other finches. But this explanation is far from complete. A male chaffinch may be happy to copulate with a female canary, but something in the female's oviduct recognizes the chaffinch ejaculate as alien and prevents his sperm from either getting near the egg or fertilizing it. The female possesses some mechanism that tells her that chaffinch sperm are not right, and on 99 per cent of occasions blocks further progress. We do not know exactly what this process might be in the canary, but studies of the chicken indicate that the block to sperm's progress lies in the vagina. If you artificially inseminate a chicken with semen from a turkey, few if any of the sperm find their way into the sperm storage tubules and at best only 2 or 3 per cent of eggs will be fertile. If, however, you were to perform a deep artificial insemination, placing the turkey semen beyond the utero-vaginal junction, you would find that 20–30 per cent of the eggs were fertilized and produce viable offspring. Similarly, if you performed another experiment and placed pieces of chicken tissue bearing sperm storage tubules together with turkey semen in a culture medium, after a few hours you would find the tubules to be full of sperm. In these latter two experiments the turkey sperm did not have to endure the vagina and made it into the storage tubules and on to the site of fertilization. This indicates that the vagina is where alien sperm are recognized and where their progress is blocked.[26]

Something very similar happens in fruitflies. If one fruitfly species inseminates another, the female may fail to lay any eggs. If you dissect the female and examine her reproductive tract under a microscope you can see that she has plenty of sperm, so her failure to reproduce it is not due to a shortage of sperm. In fact, her sperm-storage organs are likely to be full, so it looks as though the sperm are stuck in there and unable to get out. Speculating that the sperm were waiting for the right message, fruitfly biologists artificially inseminated some seminal fluid, but no sperm, from a male of the same species as the female. Lo and behold: the sperm were set free from her sperm store and the female started to lay fertilized eggs! This neat experiment elegantly demonstrates that seminal fluid from the other species simply did not contain the right messengers to trigger off the full sequence of reproductive events that lead up to fertilization.[27]

Now, if females have the physiological machinery to discriminate between the sperm of their own species and those of another, it is not too difficult to imagine that they might also have the capacity to discriminate between the sperm of different males of their own species. Once again fruitflies provide the evidence for just such an ability. In one species of fruitfly whose larvae feed on the juices of rotting cacti in the deserts of the southern United States females inseminated by their brother failed to produce fertile eggs. And, again, this was because their seminal fluid contained inappropriate signals to release the sperm trapped in the female's spermatheca.[28]

Mice provide an even more remarkable example. The rate at which sperm are transported through a female's oviduct is dependent on the match between the male and female genotypes. In this case the signal that facilitates sperm movement is unlikely to reside in the seminal fluid and more probably is provided by the sperm themselves. This is an unexpected result because for a long time sperm were thought not to express on their surface any of the genetic material they carry in their highly condensed nucleus. However, an increasing number of studies indicate that this cannot be true and that sperm must advertise something about themselves; otherwise it is difficult to see how the differential transport of sperm or the differential fusion of male and female gametes could arise.[29]

This is the process referred to in a convenient shorthand as female sperm 'choice'. This process hardly lends itself to observation since it takes place at a microscopic level and inside the female reproductive tract. For this reason it is referred to as 'cryptic female choice'.

The idea of cryptic female choice came of age in 1996 with the publication of Bill Eberhard's book *Female Control*, in which he documents dozens of instances across the entire animal kingdom in which females might exert post-copulatory control over which male or which sperm fertilizes their eggs.[30] However, long before this time, in the 1940s and 1950s, some (not many) reproductive biologists speculated that the female reproductive tract might do rather more than simply act as a conduit for sperm and eggs. Later, in the 1980s, Randy Thornhill at the University of New Mexico drew specific attention to the idea that females might copulate with several males and only after having received their sperm discriminate between them.[31] Although Thornhill's paper was a landmark, there was no intellectual stampede to see whether cryptic female choice was fact or simply a nice idea. This lack of immediate interest may have existed partly because behavioural ecologists were still busy at that time trying to demonstrate pre-copulatory female choice, but also because the 'sociological' climate was not right: we were still entrenched in a male view of sexual selection. With hindsight it seems entirely logical that under certain circumstances females would benefit from being able to control which sperm fertilized their eggs. The most likely circumstance is when females have limited control over who they copulate with. In many cases, as we have seen, females can exert considerable pre-copulatory choice over who inseminates them, but there are some instances where females are bullied or coerced into copulating. In many insects, as with the yellow dungfly, females are simply grabbed and forcibly copulated by the larger males. Much the same appears to happen in some reptiles and several ducks. In these circumstances it seems obvious that females would benefit from being able to eject ejaculates from unattractive males and retain those from preferred males. If cryptic female choice is reality and not fantasy, and if females possess the ability to use the sperm of males differentially, this creates an interesting evolutionary scenario. It means that even after copulation males do not have things all their own way, and each sex will be grappling for control over fertilization: sexual conflict again.

Bill Eberhard has championed the female perspective and identified at least twenty different ways in which females can modify or control the outcome of copulations with two or more partners.[32] Obviously, at a behavioural level, females can decide who to copulate with: this is pre-copulatory choice. Once she has started to copulate with a male she can exert post-copulatory choice in a variety of ways. She may allow a male to copulate with her but then, by failing to provide the right sensory feedback, prevent him from transferring sperm. Another possibility might be for a female to regulate the duration of copulation. In many insect species sperm transfer can take several minutes or even hours. By regulating the duration of copulation a female can manipulate how much sperm she receives from a particular male and hence the likelihood of that male fertilizing her eggs. If a female cannot regulate the duration of copulation, they might be able preferentially to retain or eject the sperm from particular males. Or a female may fail to transport the sperm of particular males to storage organs or the point of fertilization. Still another possibility is the preferential ovulation or maturation of eggs after being inseminated by particular males. Even after fertilization has taken place a female still has the potential to exert some control over her reproductive success by the differential abortion of zygotes, by investing in them differentially or even conceivably by the selective feeding or killing of neonates.

One of the most extraordinary and convincing examples of cryptic female choice occurs in a comb-jelly (technically known as a ctenophore) and known only by its scientific name: *Beroë*. I well remember seeing preserved comb-jellies in my undergraduate practical classes, mainly because they never looked anything like the illustrations in our textbook. In life *Beroë* is a beautiful bell-shaped beast, but as a museum specimen it resembles nothing more than a blob of snot, albeit a large one. Years later off Coburg Island in the Canadian Arctic I saw my first live comb-jelly. It was exquisitely beautiful: fist-sized, blood red and bearing row on row of iridescent cilia driving it slowly through the icy waters. The embryologists of old – that is, those working earlier this century – were also in love with *Beroë* because its large (1mm in diameter), uniquely transparent eggs flagrantly exposed the secrets of fertilization. Years after I had seen my first live comb-jelly I was a speaker at the Spermatology Congress in Cairns and had just given a talk to a heterogeneous audience that included andrologists, anatomists and a smattering of developmental biologists. They were titillated by the evolutionary view of sperm competition and openly sceptical of the possibility of cryptic female choice. But one developmental biologist, Gerry Schatten, came up to me afterwards and told me about two French biologists who had recently described something he thought sounded like cryptic female choice going on inside a *Beroë* egg.[33] Typically, several sperm penetrate a single egg. The female pronucleus then glides through the cytoplasm, travelling at about 0.2μm per second,* visiting each male pronucleus in turn before returning to one and fusing with it. The parallel between the pre-copulatory choice of females at a lek and what happens inside *Beroë*'s egg is striking. So here, at the level of organelles, cryptic choice appears to be happening. Unfortunately, we know little more than this. It is not known whether the sperm that an ovum appears to choose between are from the same or different males. Clearly *Beroë* provides a window of opportunity to investigate the basis of cryptic choice.

The trajectory of cryptic female choice as part of sexual selection is typical of many areas of science. An idea is

* Since there are 1000μm in 1 mm, *Beroë* pronuclei take the equivalent of 83 minutes to travel 1mm.

suggested, and ignored; subsequently the same idea is raised again, sometimes in a modified form, but still nothing happens. Then, some time later, for no obvious reason, the fuse catches, the subject suddenly becomes very popular, and everyone is doing it (see also chapter 1, p. 21). This is exactly what happened with cryptic female choice.[34] What ignited the powder trail is unclear. It may simply have been that the accretion of ideas reached a critical threshold. As we saw in chapter 1, when this happens, it can generate a tidal wave of uncritical enthusiasm and this is precisely what occurred with the study of cryptic female choice. Several papers were published in which authors, on the basis of the merest whiff of cryptic female choice, made a strong assertion for its occurrence. For example, in many species, including flies, zebra, mice, quail and humans, females eject semen some time after insemination. In several instances this ejection of sperm was assumed to constitute evidence of cryptic female choice. Clearly, if females *differentially* eject sperm, then the potential certainly exists for them to exert some post-copulatory choice over who fertilizes their eggs. But, on its own, the ejection of unidentifiable sperm is not evidence of female choice of sperm.

Golden Flies

The reproductive system of the female yellow dungfly includes two or sometimes three sperm-storage structures, the spermathecae. The existence of several physically separate sperm stores provides females with the potential to utilize sperm from their different partners differentially. Whether they do or not is another matter. However, Paul Ward at the University of Zurich conducted a series of experiments, the results of which led him to claim that females stored and used the sperm of larger males preferentially over those of smaller males. Ward allowed females to copulate with two males of different size and, in an effort to control the numbers of sperm inseminated, he standardized the duration of copulation. So after twenty minutes all copulating pairs were separated. Ward then looked at the paternity of the ensuing offspring and found that the majority was fathered by the bigger of the two males and proposed that this was because the female preferentially allowed the sperm of the larger male to fertilize her eggs. There was a certain intuitive appeal to this scenario: big males were likely to be more successful in competition for females than small males, so if a female ensured that her eggs were fertilized by a big male she would increase the chances of producing big, successful sons, and hence their success. On the face of it this all seemed entirely logical. However, Leigh Simmons and Geoff Parker in Liverpool thought otherwise.[35] Parker had studied dungflies for decades and knew that bigger males transferred sperm at faster rates than smaller males during copulation. So standardizing the duration of copulation, as Ward had done, did not control sperm numbers in the way he assumed, and in a subsequent study Simmons and Parker showed convincingly that Ward's results could easily be explained by the differences in sperm numbers. Bigger males fathered more offspring, not because females preferentially utilized their sperm, but simply because bigger males transferred sperm to females at a faster rate.

This pair of disparate studies epitomizes the problem of demonstrating the existence of cryptic female choice. After a female has been inseminated by two or more males, the sperm that result in fertilization are potentially determined by two processes: a male process (sperm competition) and a female one (sperm choice). To demonstrate either you have to control for the other. We know from the huge amount of accumulated evidence that sperm competition occurs, but are less certain about the ability of females to discriminate between the sperm of different males. In order to establish whether female sperm choice occurs we need to be able exclude the possibility that the results are due only to sperm competition.

Snakes and Lizards

Thomas Madsen is a reptile enthusiast who has managed to turn a lifelong hobby into a scientific career. More than that, some of his findings have helped to transform the way we think about female copulation strategies. Studying a population of adders in southern Sweden, Madsen and his co-workers found that females who copulated with several males had higher reproductive success than females with only a single partner.[36] Females inseminated by several males produced fewer stillborn and physically deformed baby adders and more offspring overall than monogamous females. Madsen and his colleagues reported their remarkable results in the journal *Nature* and suggested that because females could not discriminate between good and poor males on the basis of their external appearance or behaviour, instead they copulated with any male they encountered and allowed their sperm to fight it out in the female's reproductive tract. Madsen and his colleagues proposed that poor-quality males produced poor sperm; sperm with reduced competitive ability and carrying an inferior genetic message. Although a similar link between a male's quality and the quality of his sperm had been proposed for other species, such as lions,[37] many biologists were not convinced. The animal breeders who conducted their heterospermic inseminations (above) had shown time after time that a male's fertilizing capacity was independent of his behaviour or social status. Geoff Parker[38] also cast a shadow of theoretical doubt over this interpretation for Madsen's results, pointing out that geneticists know that sperm do not express the genes they are carrying so there was no known mechanism by which females could recognize and utilize superior sperm. A further, denser shadow swept in from the southern horizon. A group of Italian biologists wrote back to *Nature* reporting a complete lack of corroborating evidence from *their* population of adders: the reproductive success of females in Italy was independent of the number of sexual partners.[39]

Despite this, Madsen's idea didn't die. One of his students,

Mats Olsson, was conducting a similar study on another reptile, the sand lizard, also in southern Sweden. Remarkably, Olsson's lizard results were virtually identical to those of Madsen: the more partners a female sand lizard had, the greater her reproductive success.[40] Again it looked as though females were selecting the best sperm: cryptic female choice. Olsson worked hard to find an explanation for his results and arranged sand lizard liaisons in the laboratory to establish what determined which of two males fathered a female's offspring. Olsson used DNA fingerprinting to establish paternity, and found that one male usually fathered more offspring than the other. Unlike many other species, it appeared that male success had nothing to do with the order in which the two males copulated, and was probably also largely independent of the numbers of sperm they transferred. When he came to check the DNA fingerprints, however, Olsson and his co-workers noticed that within each family the male who fathered the more offspring was the one whose DNA profile differed most from that of the female. This held true for all their results: for any pair of males, the male who was genetically more similar to the female fathered the least offspring (figure 14).

In the light of this result, both the sand lizard and the Swedish adder studies made sense, and at the same time raised some previously unrecognized issues. The study populations of snakes and lizards in southern Sweden were small and isolated, and as a result contained many genetically similar, related individuals. Inbreeding depression is a well-known phenomenon. Darwin attributed his sickly children to the fact that his wife Emma was his cousin.[41] The fact that the Italian group, whose adders were part of a large, outbred population, failed to replicate Madsen's results was no longer a problem: their adders did not need to copulate with multiple males, and when they did so there was no obvious benefit. On the other hand, adders and sand lizards in southern Sweden copulated with several males in order to maximize the chances of finding some genetically compatible sperm.

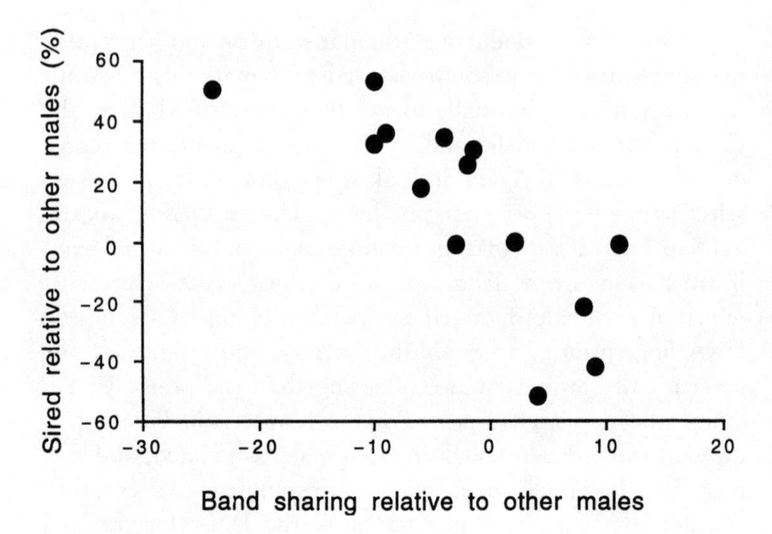

Figure 14

Sperm competition in sand lizards. The relationship between the share of paternity attained by one of two males and the degree of genetic similarity (measured as band sharing measured on DNA fingerprints) between that male and the female who produced the offspring. The more similar the male is to the female the fewer offspring he fathers (from Olsson *et al.* (1996)).

One other study, this time on a much less charismatic but no less interesting beast, the cowpea weevil, provided results consistent with those from the reptile research. Nina Wilson and her colleagues at Sunderland University arranged for several female beetles to be inseminated by the same pair of males in the same order. Half the females were completely unrelated to each other; the other half were sisters. With the unrelated females there was little or no consistency in which of the two males fathered the most baby beetles. However, with the related females, there was remarkable consistency indicating that the female's genetic make-up had an important influence on the outcome of sperm competition.[42]

Superiority Versus Compatibility

The studies by Madsen, Olsson and Wilson and their co-workers are among the very few that provide convincing evidence that females can influence the outcome of sperm competition through the choice of sperm. It is worth pointing out that there have been careful studies of other species in which researchers sought but failed to find any evidence of cryptic female choice, so it is by no means universal.[43] It is also important to emphasize that what the reptile and cowpea weevil studies do not do is provide evidence for sexual selection by means of sperm choice. In Bill Eberhard's view cryptic female choice should result in females acquiring superior sperm – that is, the sperm that all females in a population would benefit from having to fertilize their eggs. If this was the case and females agreed about what was the best sperm overall and preferentially utilized it, then there would be strong, directional selection on males and their sperm. Those males who did not produce high-quality sperm would never fertilize eggs and would soon be eliminated. Selection for good sperm by females would go on and on, producing, just as it has done with the peacock's tail, more and more extreme characteristics,[44] but so far, it does not appear that females are unanimous about what constitutes the best sperm.

Instead, these studies indicate that there is no female consensus about what constitutes the 'best' sperm, but that what is good for one female might not be any good for another: compatibility rather than superiority. Botanists have known about compatibility or complementarity effects in plants for some time, but it wasn't until relatively recently, when zoologists interested in sperm competition started to read their papers, that it became apparent that something similar might be happening in animals. Compatibility is more apparent in plants than animals. Being fixed to the spot, plants are more or less passive recipients of pollen and it therefore makes sense that they should have a mechanism to avoid being fertilized by other species, by relatives or by other genetically undesirable individuals. In contrast, most female animals are not passive recipients of sperm. They often choose whom they copulate with, and only under those circumstances where they cannot accurately discriminate between males do they resort to a post-copulatory screening process. Interestingly, and exactly as we might predict, genetic compatibility appears to be crucial in the few immobile animals, such as seasquirts, in which patterns of fertilization have been investigated.[45]

These are still early days in the field of cryptic female choice so we must be careful about drawing conclusions. However, if it proves to be generally true that compatibility matters, as it does in the adder, sand lizard and the cowpea weevil, then sexual selection via female sperm choice will be relatively unimportant. This is because females will show little or no consistency in their choice of sperm because what is good for one female may not be good for another. The important point is that the lack of female consistency will generate little skew in male reproductive success and hence cryptic female choice will not constitute an important directional force in sexual selection.[46]

Compatibility has come as a bit of shock because for a long time population geneticists have assumed most natural animal populations to be so large and so well mixed that the chance of ending up with a relative as a partner is negligible. If this is true then the adder, lizard and cowpea weevil studies have little relevance to the real world. But several studies of birds have clearly shown that the more similar the DNA fingerprints of male and female partners the higher the proportion of their eggs that fail to develop or hatch, suggesting that some degree of inbreeding does occur, and its avoidance by complementarity is crucial.[47] It might not be the only thing that matters, but it is important.

This chapter has been concerned with mechanisms – the physiological processes, including cryptic female choice, that determine the outcome of sperm competition. In the following chapter we move to another level of explanation, and consider the evolutionary significance, or, as it is also known, the adaptive significance of polyandry itself.

7 The Benefits of Polyandry

By faith and honour,
Our madams mock at us, and plainly say
Our mettle is bred out; and they will give
Their bodies to the lust of English youth
To new-store France with bastard warriors.
SHAKESPEARE, *King Henry V*, III, v, 27–31

The idea that females have anything at all to gain by copulating with more than one male has had a rocky history. With hindsight, the way sperm competition as a discipline has developed, particularly with respect to the female, seems to have been perverse. Almost from the beginning, feminists interested in sexual selection ranted and gnashed their teeth in frustration at its androcentric bias, but until recently to no avail.[1] However, few areas of science flow logically from step to step. We were wrong-footed initially by Darwin, who, as we saw in chapter 1, assumed females to be monogamous. But even a hundred years later when Bob Trivers gave sexual selection an intellectual facelift – which included the recognition that sperm competition might be important – females still got short shrift.[2] There is a certain amount of both irony and predictability in this. Trivers's brilliant insights revolutionized our ideas about sexual selection and helped to spawn the new field of behavioural ecology. But following so hard in Darwin's footprints, it is perhaps not too surprising that he should have retained a male-orientated view – after all, almost everyone else had done so. None the less, Trivers's ideas were so influential that for over a decade they perpetuated the male bias in sexual selection and in sperm competition in particular.

As we have seen, a basic assumption throughout much of the brief history of sperm competition has been that males have lots

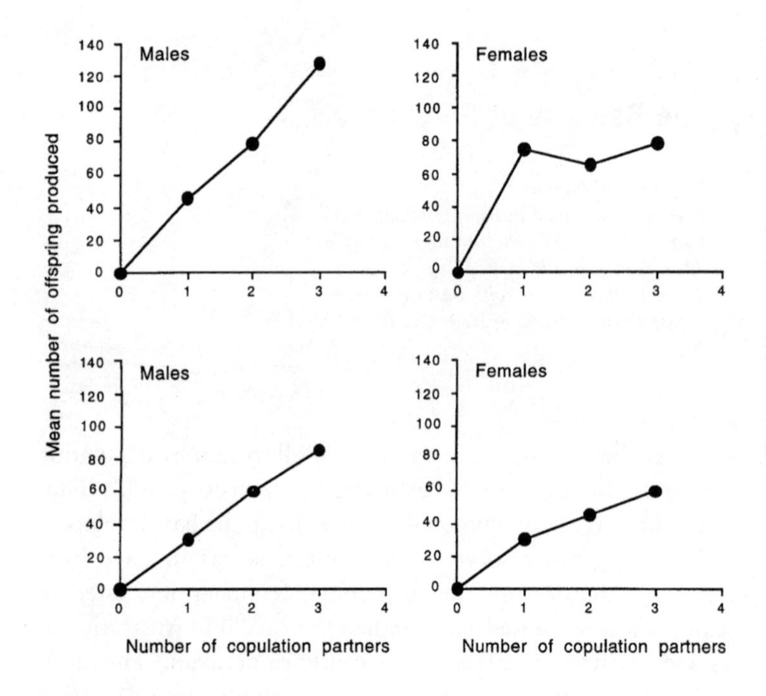

Figure 15

The full results from Bateman's experiments on *Drosophila*. The upper half of the figure is the same as figure 1. The lower half of the figure shows the results of other experiments by Bateman (1948). The upper figures show that male reproductive success increases with the number of copulation partners, but female reproductive success does not. The lower figures, derived from experiments in which food was limiting, show that female reproductive success also increased with the number of copulation partners. This result probably arose because sperm production and ejaculate size were smaller when food was limiting and females copulated repeatedly in order to obtain sufficient sperm to fertilize their eggs (from Arnold and Duvall (1994)).

to gain from copulating with multiple partners, but there is little or no benefit for females behaving similarly. This assumption arose from Trivers's interpretation of Bateman's earlier experiments on fruitflies: rampant males and acquiescent females. However, until recently it was not widely appreciated that some of Bateman's experiments did show that females benefited from copulating with several males (figure 15). Moreover, Bateman identified what this benefit was: a renewed supply of sperm.[3] Accidentally, the conditions in which Bateman maintained his flies in these particular experiments were poorer (but also probably more natural) than in the others, as reflected by their overall lower reproductive success. But an additional consequence of the poorer conditions was that males probably inseminated fewer sperm in each ejaculate. As a result, females needed to recopulate to replenish their sperm supplies sooner than they would otherwise have done. It is unfortunate that only half of Bateman's experiments showed this effect, and even more unfortunate that Trivers chose to focus on the other half.[4] The fact that Bateman got these two different sets of results was a consequence of the species of fruitfly he used and the design of his experiments.

Bateman used the species almost everyone else used at that time: the black-bellied fruitfly. Male fruitflies are ready to copulate with any female fly they meet. Virgin female fruitflies are also keen to copulate because they need a supply of sperm, but thereafter they are reluctant suitors. In part this is because males have made them reluctant by adding an anti-aphrodisiac to the semen they inseminate into them, but this effect is relatively short-lived. In general, female fruitflies copulate again only when their supply of sperm starts to dwindle. This is the key point. A single copulation will provide a female black-bellied fruitfly with enough sperm to maintain a high level of egg fertility for about four days. After this, because females don't utilize sperm especially efficiently, some of the eggs a female lays will not be fertilized. Consequently, about four days after their initial copulation females become interested in copulating again. Because Bateman's experiments ran for only three or four days, relatively few of his females needed to recopulate. Over this period, only those flies reared under the poor conditions needed to do so. Bateman therefore missed the females' renewed interest in copulating that would have occurred had his experiments gone on for longer.

There are several fruitfly species where females store so few sperm that they have to recopulate much more frequently than every three or four days to maintain the fertility of their eggs. The most extreme cases involve females recopulating after only a few hours. Had Bateman chosen a species that typically recopulates more often than every three or four days, Trivers would not have been able to disregard those results which did not fit his preconception about sexually motivated males and reluctant females. However, Trivers was not alone in this view. Coming at sexual selection from a slightly different direction, Geoff Parker arrived at a similar conclusion.[5] For males more copulations with more partners meant more offspring, but the best females could hope for from copulating with additional partners was better-quality offspring. In evolutionary terms, Parker concluded, quantity was more important than quality.

An important turning point, at least for biologists studying sperm competition in vertebrates, was a paper by Susan Smith of Mount Holyoke College, Massachusetts, published in 1988. In fourteen years of watching black-capped chickadees Smith witnessed just thirteen extra-pair copulations.[6] In nine of these the incident took place inside the extra-pair male's territory, strongly suggesting that the female had gone looking for illicit sex. Not only that, but every single extra-pair copulation involved females copulating with males that had been dominant over their partner during the previous winter. Females therefore appeared to go up-market for their extra-marital liaisons. Smith's study was pivotal in alerting us to the possibility that females might gain something from engaging in extra-pair activities. Further confirmation was provided by a group of Belgian behavioural ecologists studying another socially monogamous species, the closely related blue tit. In a paper published in *Nature* they showed that extra-pair copulations and extra-pair paternity were all down to females. The senior author of this paper, Bart Kempenaers, then at the University of Antwerp, was able to watch female blue tits during the time they were fertile.[7] His observations revealed that, far from being reluctant or passive participants in extra-pair copulations, females actually initiated them. These two studies were soon followed by others which also found females to be in control of the number of sexual partners. The myth of the reluctant female had started to crumble.

Once it was recognized that females actively seek copulations from more than one male other issues immediately became important. First, it strongly suggested that females *did* have something to gain from promiscuity. Second, if promiscuous females left more descendants than those which were not, one could infer that sexual selection on females might be much more important than had previously been thought.

We shall consider these two issues in turn and start by looking at what possible benefits females might obtain from copulating with more than one male. As will become apparent, there has been no shortage of ideas. Although inappropriate in many circumstances, thinking about why human females might have several sexual partners can give us some insights here. There are two broad possibilities. Females may trade sex for resources: money, food, a house, parental care or fertility. Alternatively, or in addition, females may engage in extra-pair copulations in order to improve the genetic quality of their offspring.

Before looking at these two classes of female benefit, we shall briefly consider (and rapidly dismiss) another possibility: that females are promiscuous simply because males are promiscuous. Just as peahens carry traces of the green iridescent plumage that renders peacocks so spectacular, females might carry some of the male's genetic baggage that controls their copulation behaviour. Put another way, polyandry may be merely a by-product of selection on males to copulate at the drop of a hat. As soon as the suggestion had been made that female copulation frequency might be merely a non-adaptive consequence of selection on males, several biologists responded by identifying studies that showed that this was not the case. A striking example is provided by chickens subject to artificial selection for either high- or low-copulation frequency. After twenty generations there was a striking change in mating frequency among cockerels, but no change in the propensity for hens to copulate, showing clearly that there was no correlation between male and female copulation frequency.[8]

The types of benefit that females might accrue from copulating with more than one male, which we have just considered for humans, can be divided into those you can see and those you can't. In behavioural ecology terminology, these are, respectively, direct benefits – those a female gets for herself – and indirect or genetic benefits – those she gets for her offspring.

The Direct Benefit of Fertility

A relative of mine is a surgeon. A couple came to see him to discuss and arrange the husband's vasectomy. They had three children, and they now needed a form of birth control. The man was subsequently booked in, but on opening up his scrotum on one side, he was found to be suffering from a congenital deformity in which the vas deferens is unattached to the testicle. It was the same on the other side. There was obviously no way in which this man could have fathered his children and clearly the woman had copulated with one or more other men in order to become pregnant. Strictly speaking, this example does not involve sperm competition because the woman's partner was 'azoospermic' and produced no sperm. This was not an isolated incident. Midwives tell of pregnant women on the maternity ward who confide in them, saying, 'I hope it doesn't look like its real dad.' More quantitatively, one study followed the fate of seventeen women married to azoospermic men and seeking fertility treatment. No fewer than four (23 per cent) of the women became pregnant before receiving any treatment and subsequently admitted to engaging in extra-marital sex.[9] As the author of this report points out, this apparently high frequency may have been partly a consequence of the National Health Service's long waiting lists for women seeking donor insemination. On the other hand, the result is exactly what one would expect if females were behaving in an adaptive manner.

There is a nice parallel to this story. In an early attempt to establish whether extra-pair copulations were adaptive in red-wing blackbirds, researchers vasectomized a number of territory-holding males. Despite the inability of these males to produce an ejaculate, their partners had no problem producing fertile eggs – with the help of the intact males next door.[10]

If you were to ask anyone on the street what the main benefit from polyandry might be for a female, most would say 'fertility'. The idea that females copulate with several different males as an insurance in case their partner is infertile is a widely held belief. As a general explanation for polyandry, the evidence for the fertility hypothesis is mixed, however.

Prior to the 1960s and 1970s many entomologists assumed that female insects could get all the sperm they required from a single copulation. Since some female insects, such as ants, were known to copulate just once and yet produce fertile eggs for years, there was abundant circumstantial evidence that made this idea credible. In addition, stuck with Darwin's female monogamy myth, many entomologists felt that females *ought* to copulate just once. However, as the study of sperm competition expanded from its beginnings in the 1970s, it became clear that the females of many insects were not monogamous and regularly copulated with several males during their lifetime. Mark Ridley found that the more polyandrous females were, the greater their reproductive success, strongly suggesting that their reproductive output was often limited by a shortage of sperm.[11] If females use repeated copulation to replenish their sperm supplies this raises other questions. Why don't males transfer enough sperm to fill a female's sperm stores? And why don't females store enough of a male's sperm to ensure a lifetime's supply?

A possible explanation for why males fail to inseminate enough sperm to keep a female going for life is that in species in which females copulate with several males it might not be worth it for a male – either because at some time in the future his sperm might be displaced by another male, or because he won't have enough sperm if he encounters another female. Putting in more than enough may simply not make economic sense for males. The cost of producing sperm is probably not trivial (chapter 3) and selection would rapidly penalize those males who were profligate and favour those who used their sperm prudently. A male's best option might be to inseminate the number of sperm that maximizes his chances of fertilizing the female's next batch of eggs, but no more. The solution to the puzzle of how much sperm a male should transfer depends on the likelihood that the female will copulate with another male and the likelihood that a subsequent insemination will displace or devalue the sperm of the original male.[12] Although this parable of the prudent male seems plausible, I think it is an unlikely explanation because in many species males inseminate many more sperm than a female can utilize. If this is generally true we are left with the second question: why don't females store enough sperm?

One answer could be that keeping sperm alive for weeks is costly for females. We know nothing about the energetics of sperm storage and we do not know how the responsibility is divided up between the sperm and the female. Nevertheless, it seems likely that the female incurs some energetic cost from keeping sperm alive inside her. If so, then it may pay females to store only as many sperm as they need in the short term, copulate when they require sperm, use it up on a current batch of eggs and then copulate again for the next batch of eggs. Another possibility is that storing a large amount of sperm may take up space within the female's body. In some insects the sperm-storage structures are fairly bulky and occupy a substantial part of the female's abdomen. The size of the storage organ will depend on the number of sperm the female needs to store, and this in turn may be determined in part by the size of individual sperm. As we have seen, fruitfly species show more variation in sperm length than any other group of animals. In the fruitflies that produce long sperm the female's storage organ, the sperm receptacle, is also much enlarged, but the larger the sperm, the fewer she is able to store and the more frequently she has to copulate. This is well illustrated by two species: the sperm of *D. acanthoptera* are just 5mm in length and a female stores about 1,400 of them which last her for about three weeks. In contrast, *D. hydei* has 23mm-long sperm, and females store just 200; to maintain their fertility females of this species must copulate five or six times every day.[13] The more frequently a female recopulates, the more likely it is that sperm competition will occur, depending, of course, on when remating takes place relative to how many sperm the female has in her sperm store. If a female waits to remate until her sperm stores are empty, then sperm competition will be non-existent. But if, as occurs in nature, she remates as soon as she's less than half full, sperm competition will be intense.

Although many female insects are polyandrous to ensure that they have enough sperm to fertilize their eggs, fertility insurance does not provide a general explanation for polyandry in other animals. The idea of a fertility benefit has been explored in most detail in birds. When I was collecting information for a book on sperm competition in birds I asked dozens of ornithologists whether they had ever come across any evidence for infertile males. There was just one instance: a kestrel, who when paired to a particular female produced a clutch in two successive seasons, both of which failed to hatch. On its own this doesn't provide evidence for male infertility. The hatching failure may just as well have been due to the female. However, in the following year the male switched partner and produced two further unhatched clutches, while his original partner successfully reared chicks with a new male.[14]

Several studies of birds, however, have proposed that fertility insurance may be one reason why female birds seek extra-pair copulations. Jon Wetton and David Parkin of Nottingham University looked at parentage in house sparrows and found that broods that included extra-pair offspring also tended to contain at least one unhatched egg. They suggested that this association between illegitimate offspring and unhatched eggs arose because the female's social partner suffered a shortage of sperm and the female sought another male to fertilize her eggs. This argument has a certain appeal but it does not fully explain the results. If the female really was copulating with another male to ensure the fertility of her eggs, her best strategy would have been to copulate early enough so that *none* of her eggs failed to be fertilized. In other words, there should have been no link between extra-pair paternity and unhatched eggs. The other problem is that it was assumed that unhatched eggs were infertile. This might not be true. If the embryo dies at a very early stage of development, say in the first one or two days after fertilization, the lack of development makes the egg appear infertile. This can be checked relatively easily, and I did this by collaborating with a Spanish biologist, José Veiga at the Zoology Museum in Madrid, who was also studying house sparrows. You can tell whether an egg is fertile or not by 'candling' it – shining a bright light through the shell: a fertile egg reveals the pink mass of the developing embryo. The eggs of most small birds, including sparrows, show signs of embryo development after two days, so we collected eggs that had been incubated for this long but showed no sign of development. We removed the layer of tissue that surrounds the yolk (see chapter 6) and, using a fluorescent dye to stain the sperm nuclei, we checked for the presence of sperm. To our surprise, about 80 per cent of all eggs had sperm present, suggesting that most unhatched eggs had been fertilized, but that the embryo had died. Obviously, this does not support the idea that female sparrows engage in extra-pair copulations for fertility purposes.[15] It does, however, suggest another possibility: the embryos could have died for a number of reasons, including the possibility that the male and female were genetically incompatible. Genetic incompatibility is well known in humans, as evidenced by the spontaneous abortion of very early embryos when both parents are of the same HLA haplotype; that is, have the same human leukocyte antigen system – part of the human version of the MHC (major histocompatibility complex). Wetton and Parkin's sparrows might therefore have copulated with other males in an effort to find a genetically compatible partner.[16]

Evidence for male infertility is more convincing in the red-winged blackbird – of vasectomy fame. There is a greater likelihood of males of this species suffering from a temporary shortage of sperm for two reasons. First, males are polygynous and may have as many as twenty females in their harem, several of which may require sperm on the same day. Second, we know that male red-winged blackbirds typically transfer about 12 million sperm during each copulation and that their sperm stores are limited to about 112 million sperm, allowing them about ten copulations per day. Elizabeth Gray of the University of Washington found that the incidence of unhatched eggs in red-winged blackbirds increased with the size a male's harem, which certainly suggests a fertility problem.[17] She also found, in contrast to the Nottingham sparrow study, that broods with extra-pair offspring had *fewer* unhatched eggs. Together these two observations provide some of the most convincing circumstantial evidence that female red-winged blackbirds copulate with other males to minimize the risk of their eggs remaining unfertilized.

A female could potentially gain a fertility benefit from copulating with additional males in one of two ways. In situations where females have no indication of whether a particular male is sterile they could copulate with others to reduce the chances of being without sperm. In a population with rather frequent male sterility, selection would rapidly favour this random form of polyandry. Alternatively, in circumstances where females can judge whether particular males are capable of fertilizing their eggs, they will be selected to avoid the no-hopers. Among socially monogamous birds, females often engage in extra-pair copulations with males that are more attractive than their regular partner and it has been suggested that they might do this to ensure an adequate supply of sperm. It is possible – indeed, it seems entirely logical – that a male with well-developed secondary sexual traits should also have well-developed testes and abundant, good-quality sperm. However, we have only to think of the occasional young male heart-throb declared sterile to realize that good sperm do not automatically follow good looks. Indeed, there is no evidence, in birds at least, that they do. Female zebra finch females are attracted to males with a red beak and a high song rate as both social and extra-pair partners, but in two separate studies there was no hint that attractive males had larger testes or better or more sperm than other males.[18]

One species where a secondary sexual trait does appear to correlate with sperm attributes is the guppy. The females of these fish choose males on the basis of their colour (preferring males with orange markings), size and the vigour with which they display. Males who displayed more (but interestingly not those with more orange markings) tended to have larger sperm reserves. Guppies have no pair bonds; the only reason for male courtship is copulation (guppies are one of the few fish with internal fertilization). The association between display and sperm numbers in guppies may arise simply because males are prepared to display only when they have an adequate supply of sperm.[19]

Another explanation for females having several copulation partners is not so much to do with ensuring sufficient sperm, but with ensuring sufficient good-quality sperm. Although females of many species can store sperm for long periods (see chapter 3), sperm don't last for ever and it is possible that even though a female may have adequate numbers to start with, the ability of the sperm to produce a viable embryo may deteriorate with time. There is good evidence for this kind of effect in birds.[20] When poultry biologists started to look in detail at just how long a hen could continue to produce fertile eggs, they found that females could occasionally produce fertile eggs three or four weeks after a single insemination. However, the longer the female had

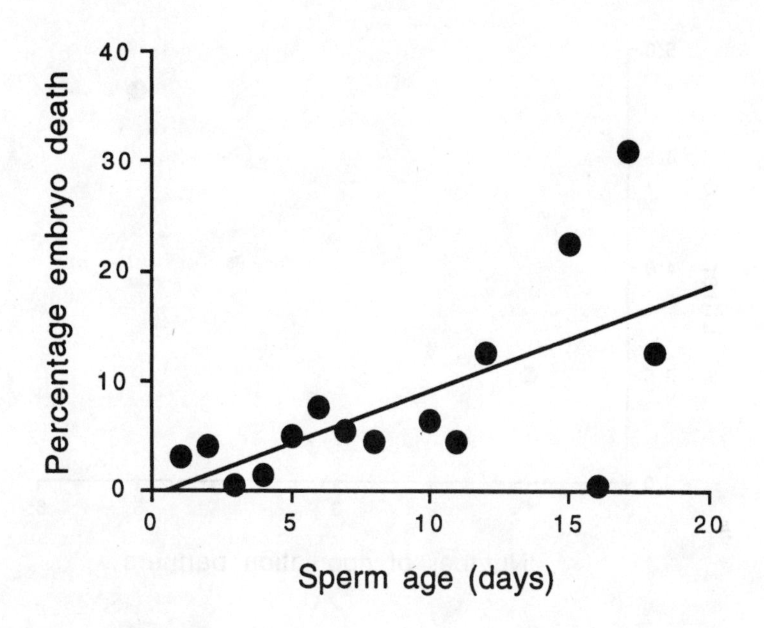

Figure 16

The relationship between the proportion of eggs with dead embryos and the age of stored sperm in the female domestic fowl reproductive tract. The longer sperm are stored before being used to fertilize the more likely the embryo is to die before hatching (from Lodge *et al.* (1971)).

stored the sperm, the greater the likelihood of the embryo dying before hatching (figure 16). The message here for hens is not to rely on stored sperm for too long. Since free-living hens copulate at least once a day, this is unlikely to be a problem.

For many years John Hoogland of the University of Maryland has studied prairie dogs, those highly social rodents of the American Midwest.[21] In several prairie dog species, including Gunnison's, females usually copulate with several males during their single afternoon of oestrus and their litters are often multiply sired. Hoogland has looked at the advantages female Gunnison's prairie dogs might gain from having several copulation partners. His results are among the most striking of any

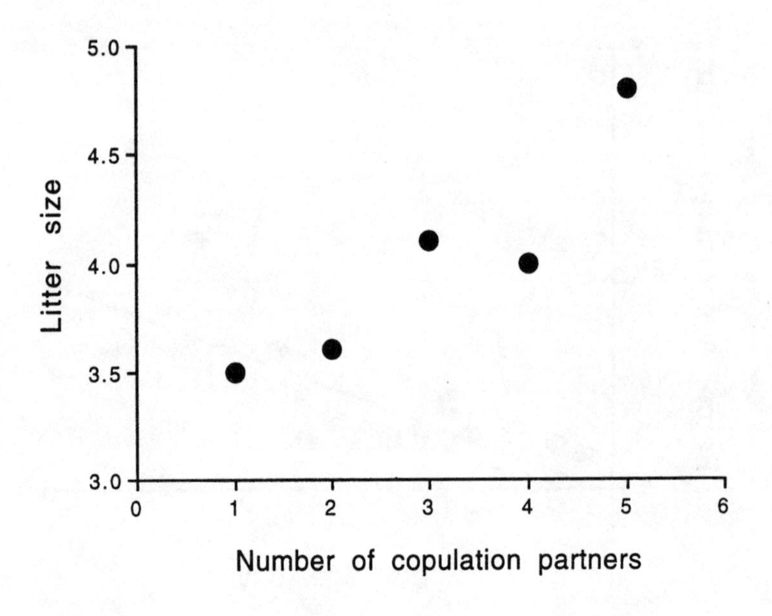

Figure 17

Promiscuity in female Gunnison's prairie dogs. The more partners a female has, the more young she produces (from Hoogland (1998)).

study so far. Females who copulated with only a single male had a 92 per cent chance of pregnancy and birth, whereas 100 per cent of those with several copulation partners became pregnant and produced a litter. This result strongly suggests that multiple partners serve as fertility insurance. The other, much more impressive result, was that females who copulated with several males produced larger litters (figure 17). Hoogland argues that it is unlikely that these females ovulated more eggs, so this effect must have some other cause. Again, genetic compatibility provides an explanation. Just as with the adders and sand lizards discussed in the last chapter, female prairie dogs with a greater number of partners may have been more likely to acquire sperm compatible with their ova, with more embryos surviving as a result.

A detailed study of a North American songbird, the dark-eyed junco, led by Ellen Ketterson of Indiana University, revealed an association between female reproductive success and the number of copulation partners.[22] Yet, despite careful observation and analysis, they were unable to find any reason why females should do better with more copulation partners. They were forced to conclude that their result might simply be an artefact of more productive females attracting more males. If this is true, the number of copulation partners a female has may be irrelevant. For a male, on the other hand, focusing on fecund females might be a very smart move. In his study of prairie dogs Hoogland was able to control for the 'better female' effect. It is well known among prairie dog biologists that heavier and older female prairie dogs produce larger litters; but even when the effects of female mass and age were controlled, females with multiple copulation partners still produced more offspring. Hoogland's study therefore suggests that females benefit from copulating with several males through a direct fertility benefit or a genetic compatibility benefit, or possibly both.

Gifts as Direct Benefits

The idea of exchanging gifts for sex is a familiar one. In humans it takes a variety of forms, some subtle, some not so subtle. Among animals, trading sex for material benefits also appears in a number of guises and, in some cases at least, can explain why females copulate more often than they need to in order to fertilize their eggs.

Most females require additional food to enable them to produce eggs or nurture their young. Most males want paternity, so they are often happy to trade food for sex. The ultimate strategy from a male's point of view would be to combine sperm with a gift in such a way that a female has to accept his sperm in order to secure the gift. Precisely this system has evolved in a number of insects in which nutrients are either dissolved in the seminal fluid or are contained in the packaging in which the sperm are delivered.[23]

For many crickets and grasshoppers copulation comprises the male attaching a large spermatophore externally to the female's genitalia. While the sperm leak out into her sperm store the female bends round and eats the wrapper – a proteinaceous coating. In some crickets the nutritious part of the spermatophore is relatively enormous, representing one-third of the male's body weight. Radio-tracers have been used to demonstrate that the nutrients in spermatophores are incorporated into a female's eggs, and it has also been found that the more a female copulates, or the more spermatophores she eats, the greater her output of offspring. The situation in some grasshopper species is extreme since the amount of nutrient a male puts into his spermatophore means that it can take him a week to produce another one. As you might expect, males are extremely fussy about who they give it to and focus their attentions on the largest, most fecund females. The corollary of this, in contrast to most other situations, is that females often have to compete for males rather than the other way round.

In some instances females may benefit from copulating with several males because they are able to digest excess sperm, exploiting the fact that males usually inseminate many more sperm than females need for fertilization. Rather than simply ejecting the leftovers, it makes sense for females to utilize the nutrients in excess sperm. Flatworms excel at few things, but they are extremely good at digesting sperm. Like many other hermaphrodites, individual flatworms exchange sperm reciprocally during copulation. However, as we have seen, this is a risky business because it is vulnerable to exploitation. One individual may accept sperm from another, but not then reciprocate. Once an individual receives a sperm donation it can use it in any way it wishes, and many flatworms treat sex as nothing more than a free meal. Since food is often in short supply for flatworms, using another flatworm's sperm is a good ploy. However, as soon as flatworms start doing this, selection will favour any individual with a counter-ploy. Some flatworms have responded by either delivering their sperm wrapped in a digestion-proof spermatophore or including chemicals in their seminal fluid that neutralize the recipient's digestive enzymes.[24]

Unconventional nuptial gifts are widespread in insects. Several species have specialized glands that produce nutritious saliva or other valuable substances on which females feed. Others have taken this even further and in certain crickets, for example, the males possess special, fleshy hind-wings which the female consumes during copulation. The ultimate nutritional gift a male can provide to a female is himself. Sexual cannibalism occurs in a range of species but is best known in spiders and praying mantis. Discussing the situation in the European praying mantis in the 1890s, J. H. Fabre wrote, 'If the poor fellow is loved by his lady as the vivifier of her ovaries, he is also loved as a piece of highly flavoured game.' Ever since, the idea of sexual cannibalism has excited speculation because it was difficult to imagine how anything so apparently detrimental could have evolved. It was therefore assumed that what Fabre had so colourfully described was an artefact of observing mantis in captivity. However, a selfish-gene view of sexual cannibalism provides a much more compelling explanation. First, whether something is good for the species is irrelevant; what matters is the individual. Second, although sexual cannibalism may be beneficial to females, males might not actually sacrifice themselves.[25]

With these ideas in her head, Susan Lawrence of Sheffield University set off to Coimbra in Portugal to be one of the first people to look at what mantis get up to in the wild.[26] She found that in about one-third of all copulations the male was killed and eaten by the female. She also discovered that, far from sacrificing themselves, males did everything in their power to avoid being eaten. This was blatantly obvious from watching copulatory sequences in nature. A couple of weeks after her final moult the female mantis is ready to copulate. She signals her readiness by releasing a pheromone and on catching a whiff of this a male will fly upwind until he locates her. However, he does not come winging in to alight beside her: this is far too dangerous. Instead, he usually alights about thirty or forty centimetres away. Both sexes have big eyes and excellent vision so the male waits until he can see the female. He then stalks her – extremely carefully. His behaviour is remarkable: if she turns to look in his direction he freezes; when she turns away he moves forward. If there's a breeze, he edges forward using the movement of the fluttering vegetation to disguise his progress. Once he is within ten to twenty centimetres of the female he waits. He composes himself and when the female is looking elsewhere, he leaps on to her back in a single adroit movement. If he does it right, he's safe. But if he misjudges his timing he ends up in her prickly embrace. Let's assume our male got it right. Grasping the female and lying longitudinally along the female's back, the male's prehensile abdomen tip starts to seek out hers, and copulation takes place. After copulating for about two hours the male either jumps or flies off the female's back as quickly as you can imagine. Again, if he messes this up, he's dead, but usually what happens is that he springs away from the female and hits the ground running. He does not want to be dinner.

Now let's consider scenario two, in which the approaching male mistimes his leap and is caught by the female. She starts to consume him straight away and, depending on how the male was caught, she may or may not start with his head. If she eats his head she also consumes the nervous tissue that controls his copulation behaviour. Having lost this particular ganglion the male also loses all inhibitions about sex and his abdominal tip goes into overdrive seeking out the female's genital opening. Once connected he is perfectly capable of transferring his sperm – despite being headless. So, males don't want to be eaten, but sometimes they are. If they are, then their reproductive end can still function reasonably well and they have a good chance of inseminating the female. On the other hand, females almost always want to eat their partner, since he represents a good meal which enables her to produce more eggs. Even if he is eaten, providing the female does not copulate with another male, he has a good chance of fertilizing her eggs. However, females can and do copulate with several males, so the male has no guarantee of paternity.

A fundamentally different cannibalistic situation occurs in the redback spider. The consumption of the male by the female during copulation is routine in this species. But, in contrast to the mantis, there is no sexual conflict here, since the male wants to die – and in a most unusual way. Within seconds of starting to inseminate the female the male redback does a somersault, landing directly on the female's mouth. Alice herself would have been hard pressed to find a more obvious signal to 'Eat Me' and in the majority of cases this is exactly what happens. So, what's in it for the male? There are two possibilities. His body may, as in the praying mantis, contribute towards the quality or quantity of offspring produced. Or his complicity in cannibalism may increase his certainty of paternity. This could be important because female redbacks often copulate with more than one male. Working in Australia, Maydianne Andrade of the University of Toronto showed that male sacrifice is unlikely to result in increased offspring quality or quantity.[27] As in many spiders the male redback is much smaller than the female, just 1 per cent of the female's body mass, and only 2.5 per cent of the egg mass she later produces, so eating him would provide little extra nourishment. But in terms of paternity there were two advantages to being eaten. Copulations involving cannibalism last almost twice as long (25 minutes) as those that don't (11 minutes), and the longer a copulation lasts the more likely the female is to use that male's sperm to fertilize her eggs. In addition, following a cannibalistic copulation, females were much less eager to copulate with other males, thereby reducing the likelihood of sperm competition. In other words, in this instance it really seems to be adaptive for males to be eaten!

The males of many bird species, such as terns and eagles,

bring food to their female partner during the days before she lays. Depending on the species, the male may provide nearly all the female's food requirements, and sometimes the delivery of a food item is followed by copulation. It seems logical that a female might be able to acquire even more food by accepting gifts from other males, in exchange for copulation. Surprisingly, female birds rarely seem to do this. An antipodean seabird, the red-billed gull, does it occasionally, as do some birds of prey, but these are the only known instances. Moreover, these transactions have a hidden meaning: females that accept food from strangers are usually those whose own partner is less than adequate. A poor provider spells doom for a female: the quality and quantity of eggs she produces will be reduced and any chicks that hatch will not be adequately provided for. Under these circumstances the female's best bet is to find someone better, and in these species extra-pair copulation may be the prelude to switching partners.[28]

The bonobo is remarkable among primates for several reasons and not least, as we have seen, because it substitutes sex for aggression. Females are receptive virtually throughout their cycle, and sex between group members, regardless of age or gender, is frequent. In bonobo society various forms of sex serve to dissipate tension, and sexual behaviour is especially frequent when there is competition for food. Part of the increase in sexual activity in response to food occurs because some females occasionally solicit copulations from males in exchange for food.[29]

Finally, in this section, there is one instance in birds where females seem to trade sex for an important resource other than food. Like most long-lived birds, Adelie penguins are socially monogamous. Using chunky pebbles, they build a raised nest to protect their eggs from melt water and the other filth that accumulates in seabird colonies. Stones are in short supply, because they are usually already part of a bird's nest. When owners are away both male and female penguins steal stones from other nests. But if a male is alone at a nest, a female from another pair may take a stone in exchange for copulation with the male nest-owner. This remarkable behaviour was discovered by Fiona Hunter, then at Cambridge University, working in the Antarctic, and when the media got hold of it, the headline was 'Penguin Prostitutes'.[30] A moment's thought, however, would have revealed that the female penguin's behaviour differs from human prostitution in several important ways. First, in humans it is usually males that seek out females. Second, males use prostitutes for sexual gratification, not to increase their chances of paternity. The fact that female Adelies can be so casual about copulating with other males is remarkable and has several implications. It could mean that female Adelies don't care who the father of their offspring is. It could mean that females have exquisite control of sperm and can eject it after insemination. Or it could be that females choose their extra-pair partners rather carefully, and get the best of both worlds: stones *and* sperm.

The Direct Benefit of Parental Care

A third, popular explanation for why females copulate with different males is to trade sex for parental care. If males are prepared to provide care for the offspring of females with whom they have copulated, females can benefit from multiple copulation partners. The benefit to such females is increased reproductive success because offspring cared for by more than one male may have a greater chance of survival. Parental care is most prevalent in birds and mammals, and can take a number of different forms. The most obvious is the provision of food, but males can also protect offspring from predators.

Polyandrous marriages in humans are extremely rare and occur only in cultures where, for ecological or social reasons, resources are extremely limited. In most instances polyandrous marriages involve two or more brothers sharing a single wife. From a male perspective a polyandrous marriage appears to be a poor deal since sharing a wife obviously limits male reproductive success. It was probably this chauvinistic standpoint that led Joseph Hooker disparagingly to liken polyandrous people in Bhutan to primitive animals. Early anthropologists perpetuated this view by referring to them as survivors from a more primitive epoch. It is rather disappointing that Hooker, Darwin's closest ally and confidant, should have failed to see that polyandrous marriage might have had an adaptive explanation – even from a male perspective. Certainly from a female standpoint the parental care, love and attention of several husbands seems very advantageous. This is verified by the fact that in Ladakh the reproductive success of women in polyandrous marriages, with an average of 5.2 children, was higher than those in monogamous marriages, who produced only 3.1 children. The social and evolutionary advantage of polyandrous marriage for males is that when resources are limited, as they are among feudal societies in the depauperate valleys of the Himalayas, two or more males are capable of supporting a wife whereas a single man cannot.[31] Thus, under difficult ecological circumstances, polyandrous marriage is a case of men making the best of a bad job. Moreover, if a male has to share his paternity with another man, in evolutionary terms he is better off doing so with a close relative rather than with a non-relative. The genetic costs of sharing paternity are offset to a considerable extent if co-husbands are relatives. Because brothers share half their genes, by helping to rear a brother's offspring, a male perpetuates some of his own genes. None the less, since any male would do even better if he were the sole husband, conflict between males in polyandrous marriages can be high. This is clearly illustrated by a natural experiment in Sri Lanka where co-husbands in polyandrous marriages may be either brothers or non-relatives. Among the former, marriages are more stable and more enduring than those involving unrelated males.

In most polyandrous marriages the oldest brother is typically a few years younger than his wife. This in turn means that the wife is slightly, or in some cases considerably, older than her husband's younger brothers, her co-husbands. The consequence of this disparity in ages is very different reproductive success for the different males. Among the Nyinba, a Tibetan group living in north-western Nepal, where co-husbands are brothers and where wives place great importance on knowing paternity, the reproductive success of the eldest brothers was greater than second and third brothers. Not surprisingly, as soon as their financial situation permits it, younger brothers leave to strike out on their own.

Competition for sexual access to the wife and sperm competition would appear to be almost inevitable in polyandrous marriages. However, in some societies, such as the Ladakhi Buddhists studied by John Crook, conflict is avoided by a clever cultural ploy. The brothers most likely to compete for paternity are the senior husband and his closest brother. To minimize potential disputes, second sons are honoured by being chosen as monks. Being selected as a monk is a sufficiently great honour for second sons to comply willingly – even though their reproductive success will be zero. For this reason, despite the apparent similarity between polyandrous marriage in humans and polyandrous mating systems in birds, the potential for sperm competition among humans is probably rather limited.

A more specialized form of polyandrous marriage is spouse exchange or co-marriage. In the early days of Arctic exploration the Inuit were infamous for allowing southern fur-trappers, whalers and traders to copulate with their wives. Inuit men would make their wife available to other men as a form of hospitality, especially if visitors were wealthy, as southerners often appeared to be. To those early explorers Inuit morals were merely lax, but a godsend for the likes of Peter Freuchen and Knud Ramussen so far from their own wives and girlfriends. Rather than being simply immoral, however, spouse exchange was an important part of Inuit culture, with clear adaptive value. Most cases of spouse exchange were actually co-marriages: the enduring co-operative and sometimes sexual relationship between two Inuit couples – a form of long-distance reciprocity with immense survival value. Co-marriage instantly increased each participant's circle of relatives, and as a result his or her likelihood of survival. Arctic conditions are often harsh, hunting success unpredictable and starvation all too familiar. Because family members were morally obliged to care for each other, having others that one could depend on must have more than offset any loss of paternity. But, if having additional family was so crucial, why didn't the Inuit exchange their spouses more widely? What was so special about reciprocating *pairs* of couples? The answer to this question has come not from the study of people but from research on the co-operative polyandrous mating systems in birds.[32]

During his brief but seminal visit to the Galapagos Islands Darwin was impressed by the tameness of the birds and in a letter home he described how he had pushed a hawk off a branch with the muzzle of his gun. Darwin did not know it, but the Galapagos hawk has a mating arrangement with females blatantly trading sex for paternal care. The mating system of these birds is extremely unusual and is described as co-operative polyandry. It comprises two or three males paired to a single female. All the males copulate with the female, often in rapid succession, and subsequently all work together to rear the offspring. Paternity analyses conducted in the early 1990s revealed that each male in a group had a similar chance of being a father, hence explaining why it is worthwhile for each of them to help rear the chicks.[33]

A similar system occurs in the dunnock, and Nicholas Davies elegantly demonstrated the benefits females accrue by having two partners. By making sure both males copulate with her, the female trades sex for paternal care, and in so doing produces more offspring and increases her own reproductive success. This looks like a case of the more the merrier for the female. But when Davies and Ian Hartley cleverly arranged things so that female dunnocks could take a third husband, they declined. The reason was that males adjust how much effort they put into feeding the chicks in relation to their share of paternity. In a group comprising three husbands each male would have a reduced chance of being a father and so would reduce his chick-feeding effort accordingly, resulting in an overall reduction in the amount of male care. For females two husbands were better than one and much better than three. When Hartley and Davies looked at the other bird species whose mating system was co-operative polyandry they found a similar pattern. In almost all cases the number of males associated with one female was rarely more than two. With more than two males in a group a male's chance of paternity is so reduced that it is not worth helping to rear offspring. Exactly the same explanation applies to co-marriage among the Inuit. Because Inuit co-families usually live in different communities, fathers provide parental care primarily for the offspring produced by their main wife. From the primary husband's perspective, his chances of being the genetic father of the offspring he helps to rear are relatively high. However, if the system of spouse exchange involved more than two couples, each male would be much less certain of his paternity and the value of his parental care would be correspondingly reduced.[34]

We can now ask whether the provision of paternal care by extra-pair males explains the high incidence of extra-pair paternity we find in socially monogamous birds. The answer is that, under some circumstances, it might. In her study of red-winged blackbirds Elizabeth Gray found that females who had engaged in extra-pair copulations with neighbouring males were less likely to lose their chicks to predators.[35] She also showed that when she placed a stuffed magpie – which in real life is a potential predator of blackbird chicks – near a female's nest containing chicks, only her partner and those males from adjacent territories with whom she had copulated came to attack the predator. This looks like a clear case of females trading paternity for protection from predators. Unfortunately, this is unlikely to provide a general explanation for extra-pair copulation behaviour in socially monogamous birds. In another study of parentage in red-winged blackbirds, females with illegitimate young in their nest had reduced rather than enhanced breeding success and in other bird species there is no evidence that extra-pair males provide any kind of parental care.

A somewhat perverse form of paternal care is *not* being aggressive to your offspring.[36] Male primates and lions are renowned for being unpleasant to babies, but only if they are not their own. One way females can avoid this abuse of their offspring is to copulate with several males and persuade each of them that they might be the father. In that way it would never be in a male's interest to injure her offspring. Polyandry may be beneficial because it confuses paternity, which may in turn increase the chances of copulating males caring for offspring, and simultaneously minimize the likelihood of infant abuse or infanticide by males. Although there has been no formal test of the avoiding-infant-abuse hypothesis, several observations are consistent with it. The most likely perpetrators of infant abuse or infanticide are males that have recently joined a group or that might take over a group – extra-group males. Females seem to be particularly keen to solicit copulations from such males. Male troop take-overs and subsequent infanticide are regular events among langurs and females will even copulate with extra-group males while a take-over is in progress. More striking still, after a take-over, females will solicit copulations from the new males, even though they are already pregnant!

Some of the most dramatic evidence that certainty of paternity influences the way males behave towards juveniles comes from our own species. When a woman rears a family consisting of offspring from both a previous partner and a current partner, the conflict between stepfather and stepchildren is considerably greater than that between biological parent and offspring. In the 1980s Martin Daly and Margo Wilson showed that stepchildren were much more likely to suffer from abuse than were children with both biological parents, and they suggested that child abuse is a side-effect of the psychological mechanisms that dictate how males should allocate care to offspring.[37] You might be forgiven for thinking that human females are unlikely to trade sex for paternal care in contemporary Western society (although I suspect we can all think of isolated examples), but in the Ache, a group of preliterate South American Indians, there is clear evidence for such an effect.

The first contact between the Ache people and white anthropologists occurred as recently as 1972 – with the inevitable but disastrous result that 40 per cent of them died from Western diseases. In reassembling a picture of Ache life before 1972, anthropologists Kim Hill and and Magdalena Hurtado of the University of New Mexico discovered an extremely brutal society in which male aggression ruled; club fights between men were common and as a result children were often left fatherless.[38] Fathers sometimes also abandoned their partner and offspring of their own accord – presumably seeking better reproductive opportunities elsewhere. Children without fathers had little better than a 50 per cent chance of survival, compared with an 86 per cent chance with their father present. In this tough hunter-gatherer society pragmatism rode roughshod over compassion, and abandoned children were usually buried alive – unless they had a secondary father. There is no marriage as such in Ache society, and women are typically polyandrous, with at least two partners: a primary partner – the one most likely to be the father of a particular child – and a secondary one, who still has some likelihood, albeit rather less, of being that child's father. Through this system of multiple sexual partners, females generated uncertainty about paternity, which ensured that if the primary father died, there would be someone else to care for and feed their children: a clear case of human females trading sex for paternal investment.

The Direct Benefit of Avoiding Rejection Costs

On the outskirts of Sheffield where I live there are numerous ponds or dams which in past times provided the power for the early steel industry. The ponds are occupied by numerous mallard ducks and each spring we are forced to witness the sexual subjugation of females by males. Forced copulation is a ubiquitous feature of mallard reproduction: fleeing females are followed by their partner and one or more other males.[39] Eventually the female is forced, exhausted, out of the air, sometimes into the middle of the school playground, sometimes back on to the pond. She is then forcibly copulated while her partner fights valiantly to remove the extra-pair male. It is brutal, brutish and extremely dangerous for females, several of which die each year as a result. But, to judge from the molecular evidence, forced extra-pair copulation is a successful strategy for male mallards.[40]

One option the female has under these circumstances is to give in to the male and allow him to copulate so that she will be left in peace. This explanation assumes that copulating with a harassing male carries no costs. This seems unlikely because often the males that resort to coercive tactics are defective in some way: low status, poor quality, or diseased and with little chance of securing a partner using more conventional tactics. If this is true then simply giving in to a male to save time and harassment does not sound like a good idea. To put it another way, this kind of female strategy would not be evolutionarily stable. If all females simply rolled over and acquiesced every time a male made a nuisance of himself, there would be chaos since all males would be at it. Whenever copulations are costly for females, for whatever reason, it will usually pay them to resist. Of course, the corollary of this is that there will be selection on males to make an ever greater nuisance of themselves. By continuing to resist, females may be trying to make persistence disproportionately costly in terms of time and energy for males.[41] If you watch mallard ducks in early spring as they are chased and harassed by male after male, this cat and mouse evolution scenario seems plausible. By far the most effective solution for females might be to accept males' advances, but then to dump their sperm. If they could do this, some of the costs at least would be eliminated, and because males would stand to gain nothing in terms of paternity, there would be no evolutionary incentive for the behaviour to persist. The fact that coercive male behaviour is widespread, and in mallards at least is quite often successful, suggests that females do not have total control over the fate of inseminated sperm.

Indirect Benefits

When females do not appear to obtain direct benefits for themselves from copulating with more than one male, the possibility exists that they might acquire indirect benefits: that is, genetic benefits for their offspring. At first sight the idea of genetic benefits seems entirely plausible, but in terms of both theory and reality it is a quagmire.[42]

There are four possible types of genetic benefit. (1) By having several sexual partners females might increase the genetic diversity of their offspring. In a rapidly changing world this might ensure that at least some of a female's offspring are adapted to prevailing environmental conditions and so will survive to reproduce. This seems sensible, but population geneticists consider that the additional genetic variation a female would gain from being inseminated by more than one male would be trivial. The point can be most clearly seen by thinking of ourselves: we each have 23 pairs of chromosomes and, as a result of meiosis and recombination, a man can potentially produce 2^{23} (i.e. an astronomical number of) different sperm types. It has therefore been assumed that by copulating with several partners females will do little to increase the genetic diversity of their offspring.[43] The result is that this particular genetic benefit has received little empirical attention.

(2) As was suggested by Trivers in 1972, females might base their choice of partner on genetic complementarity. At first no one bothered to follow up Trivers's idea because no one could imagine how, on the basis of a male's appearance, a female could assess whether his genotype would match her own. Only later, when it was clear that females routinely copulated with several males *and* that some females at least had the ability to

discriminate between the sperm of different males, as we saw in chapter 6, did the idea of post-insemination choice based on compatibility seem feasible. Jeanne Zeh of Rice University, Texas, and her husband David Zeh, now at the University of Reno, have championed the view that females copulate with different males to minimize their chances of being fertilized by an incompatible male. They point out that there is abundant evidence for genetic incompatibility – inbreeding depression being the most obvious example. They also draw attention to a recently discovered cause of incompatibility – the existence of genetic conflicts within and between the nucleus and cytoplasm. Such intragenomic conflict results in genomes being dynamic and evolving entities. This state of genomic flux will of course increase the likelihood that your genes will not be compatible with those of your partner.[44]

We now come to two genetic benefits which have been the major focus of both theory and practical research for the last twenty years. These two ideas have also been the focus of considerable controversy, partly because biologists have been divided over whether they might work, and partly because, despite this, several studies claim to have shown that they exist. They comprise two processes. The first of these (3) was proposed by the geneticist and statistician R. A. Fisher in the 1930s and is concerned with genes for attractiveness. Fisher proposed that females choose attractive males, not because these males are better in any way, but simply because they are attractive. He proposed that if genes for a female preference for particular male characteristics were linked to the attractive trait in males, a runaway process would occur. In this scenario females always choose the most attractive males and produce attractive sons so that over successive generations there is very strong directional selection which – in the most extreme cases – results in the dramatic sexual dimorphism we see in species such as birds of paradise. By choosing to copulate with an attractive male a female will produce attractive sons, who in turn will be successful in reproduction.[45]

The final type of genetic benefit (4) is viability.[46] The idea here is that male secondary sexual traits, such as shiny plumage, a brightly coloured bill, or a complicated song, are not arbitrary features, but have evolved to reflect a male's inherent quality. In this scenario attractive males are also extremely viable males. By having her eggs fertilized by an attractive male a female acquires viability genes for her offspring – genes that will help her offspring survive to reproduce.

These latter two classes of benefit have become the focus of a phenomenon known as the 'paradox of the lek'.[47] The lek mating system of many animals, including some fish, frogs, birds and mammals, in which males congregate and females turn up simply to copulate with one of the males, is a paradox because biologists cannot see what females get from it. In most lekking species the majority of females copulate with only a handful of males, so the distribution of copulations among males is very uneven. After a female has been inseminated she leaves the lek to rear her offspring alone. It is generally assumed that she gets no direct benefits from her choice of partner at the lek, and it is therefore inferred that she must obtain some indirect benefit from choosing a particular male, and the benefit is generally assumed to be genes – either for attractiveness or for viability.

But there is a major problem with the idea that female choice on leks is driven by either of these genetic benefits. If females all agree about what sort of male they want to copulate with and females always copulate with this type of male, it creates incredibly strong selection on males. Any male that doesn't conform to the female's ideal is genetically dead. He's not chosen, never copulates and the genes he carries go nowhere. The problem with directional selection is that very soon all the genetic variation in male attractiveness is used up. Breeders of domestic animals soon found this out when they put their breeding programmes into practice – they imposed strong directional artificial selection on cattle to improve milk yield, or on chickens to improve egg production. But within a few generations all individuals were much the same and no further improvement could be made, simply because all the genetic variation for these traits had been used up. If this were true of sexual selection among wild animals, then there would be no point in females exerting any choice, but they do!

One solution to this conundrum, suggested by Bill Hamilton and Marlene Zuk, is a system in which selection is not consistently in the same direction. Hamilton and Zuk proposed that parasites, by continuously evolving new varieties (and at faster rates than their hosts) might provide just this kind of regime.[48] They suggested that males infested with parasites are incapable of producing secondary sexual traits to the same extent as healthy males. By choosing males with the most elaborate traits, females acquire viability genes, specifically genes for disease resistance, for their offspring. The ingenious aspect of Hamilton's and Zuk's idea hinges on the fact that disease organisms are constantly evolving, and the consequence of this is that selection for disease resistance via female choice never continues long enough in the same direction to use up all the genetic variation.

Testing these ideas has been extremely problematic. Demonstrating indirect benefits at all has been difficult, and distinguishing between genetic benefits for attractiveness and viability has been all but impossible. There are two problems to overcome. First, researchers cannot start to consider genes for attractiveness until they have eliminated the possibility that females are gaining genes for viability and vice versa. Second, among wild animals, especially those on leks, it is extremely difficult to conduct experiments that would allow researchers to tackle the idea of indirect benefits. Ideally, what one would like to do is to see what happens when females *cannot* choose, but since females on leks always choose and it is difficult to make them copulate with another male, this is something of a dead end. We know from a number of laboratory studies of more tractable species, such as fruitflies, that if you allow a female to choose her partner her reproductive success is greater than if her partner is allocated to her at random by the experimenter. Lekking sage grouse or Uganda kob hardly lend themselves to this type of manipulation and so it looks as though the paradox of the lek may remain unresolved.

There is a feasible alternative, however. It does not involve lekking species, but comprises another situation where females may obtain indirect benefits from copulating with particular males. Imagine a female swallow migrating back from Africa to breed in Europe. She is heading back to the farmyard where she bred in the previous year, but this time she is late, delayed by a sandstorm over the Sahara. When she arrives almost all the other females are paired, and paired to the most attractive males. She has to make do with Mr Average. She pairs up with him and they build a nest. However, in the middle of her fertile period she gives her partner the slip, finds one of the really attractive males, and copulates with him. She then returns to her nest as though nothing has happened. Pairing up with Mr Average has provided the female with the direct benefit of an opportunity to breed at all, but by surreptitiously copulating at the optimal time for fertilization with Mr Much-better-than-average she can be fairly sure at least some of her chicks are fathered by the better male. So here we have a situation where a female is producing two types of baby: legitimate offspring from her social partner whom she doesn't really rate, and illegitimate offspring from her extra-pair partner whom she specifically chose to copulate with. Suddenly we have a sperm competition situation in which it might be possible to see whether females gain genetic benefits from copulating with particular males.

On the basis of this scenario we can make several predictions. If a female gets viability genes from her extra-pair partner, then her illegitimate offspring should survive better than her legitimate offspring. If there is no difference in the viability of the offspring from her different partners we can forget this particular genetic benefit and look at genes for attractiveness. The prediction in this situation is that, when they grow up, the illegitimate offspring will be more attractive to females than the legitimate offspring.[49]

This brings us to the question of what genes could confer viability benefits. The answer seems to be, for some species at least, those that provide protection from pathogens. Some of the most convincing results come from a study of the MHC – the major histocompatibility complex – of house mice. The MHC comprises a closely linked set of genes in vertebrates associated with disease resistance and the immune system. Wayne Potts from the University of Florida and his colleagues found that female mice living in the territory of a male whose MHC was similar to their own were more likely to engage in copulations with other males, especially those with a different MHC.[50] A plausible explanation for this female mate choice is that offspring fathered by a male whose MHC is different from their mother's acquire a more varied set of MHC genes and hence a more effective pathogen resistance system. Since mice can distinguish different MHC strains on the basis of urinary odours, it seems likely that females 'know' when to seek extra-pair copulations simply on the basis of their partner's smell. It remains to be shown in this system that the offspring of mice with compatible MHCs are more viable than those whose MHCs are incompatible. However, this seems to be a very clear example of a case where females can decide on the genetic compatibility of their partner simply on the basis of his scent. There is some tantalizing, but controversial, evidence from a study by Claus Wedekind that MHC may also affect choice of partners in humans, this time mediated by body odours. What is not known is whether female choice for males with particular MHC attributes is simply to avoid the general problem of inbreeding, or whether this is part of the process of sexual selection in which females acquire specific genetic benefits for their offspring.[51]

Birds are renowned for their lack of any sense of smell. None the less there have been several studies that suggest that females might gain viability benefits in the form of disease-resistance genes via their choice of extra-pair partner. In Anders Møller's now famous study of swallows, females preferentially pair with and specifically seek extra-pair copulations with long-tailed individuals. As a consequence, the reproductive success of long-tailed males is much greater than that of other males since they father offspring both in their own nest and in the nests of other males. Females also appear to benefit, this time indirectly, because long-tailed males appear to have a better-developed immune system, indicating that the reason offspring fathered by long-tailed males survive better is because they resist pathogens more effectively than those fathered by other males.[52]

In another study, the indirect benefits females accrue from infidelity appear to be genes for attractiveness rather than viability. In the scrubby regions of Australia lives a handful of tiny jewel-like wren species whose superlative names – splendid, lovely and superb – describe the appearance of the males. The superb fairy wrens breeding in the Canberra Botanical Garden have been the subject of a detailed study over the past decade by a team led by Andrew Cockburn of Australia's National University.[53] These birds live in co-operative groups which comprise a breeding pair aided by several grown sons. Astonishingly, over two-thirds of all offspring are fathered by extra-group males. During the breeding season males are like persistent salesmen relentlessly touting their wares in front of other males' females. They do this by plucking a yellow flower petal, which contrasts perfectly with their iridescent cobalt-blue plumage, and presenting this to a female. These displays eventually lead to extra-pair copulations when females surreptitiously slip away from home in the dim light of dawn to visit one of these males in his own territory. Extra-pair males provide no direct benefits to females but the fact that most extra-pair fertilizations are achieved by a dynasty of males, stemming from one particularly successful individual, strongly suggests that the benefits females accrue from extra-pair copulations are genetic. Cockburn and his colleagues were able to eliminate the idea that these genetic benefits were for viability, since there was no evidence whatsoever that extra-pair offspring had any greater reproductive success or were more likely to survive than other offspring, leaving researchers with the possibility of attractiveness genes.

What can we conclude from all this? In those species or instances where females choose to copulate with more than one male it seems very likely that they benefit from doing so. The evidence would seem particularly compelling if there exists a positive association between the number of partners a female has and her reproductive success. However, we should bear in mind that this association could arise even when females gain no benefit from multiple copulation partners – simply because high-quality females attract more males. Fortunately, in some instances it is clear that it is advantageous for females to have multiple partners – obtaining direct benefits, such as a fresh supply of sperm as in fruitflies, or parental care as in dunnocks.

Although genetic benefits continue to be controversial – because the mechanisms by which they might operate are still poorly understood – several studies have produced convincing evidence that females produce better-quality offspring by copulating with several different males. There is less evidence for the idea that females acquire genes for attractiveness through their choice of copulation partner, but this idea is even more difficult to test.

Finally, there is one further type of genetic benefit that avoids the mechanistic complexities associated with viability and attractiveness genes, and that is Jeanne and David Zeh's idea that females choose their copulation partners, or their sperm, on the basis of their genetic complementarity.[54] Several studies of pre-copulatory choice and cryptic female choice in particular have provided tantalizing evidence which is consistent with the idea of compatibility. In some instances females may be able to discriminate between males if they advertise which particular 'type' they are – as in mice, and possibly humans, which signal their MHC type via their odour. However, in the absence of such signals females may do best by copulating with several different males and allowing their reproductive tract to sort out compatible from incompatible sperm.

There is now some compelling evidence for the genetic compatibility idea, and it comes from an unlikely study organism – a tiny pseudoscorpion which inhabits rotting wood in Panamanian forests.[55] This animal is unusual in a number of ways: the first is that it gets to new rotting stumps by hitching a lift under the wings of a giant harlequin beetle. The second is its revealing mode of reproduction. Uniquely, the pseudoscorpion lays its eggs into a translucent external brood sac – effectively an external womb – enabling researchers to monitor the hatching and survival of the babies as they grow, something that is impossible in other animals which produce live young. The researchers found that female pseudoscorpions that had been inseminated by only a single male aborted a much higher proportion of their developing embryos than females that had been allowed to copulate with several partners. Cleverly designed experiments in which females obtained the same numbers of sperm from different partners revealed that abortions occurred because of genetic incompatibility between certain partners. Polyandry, therefore, was beneficial to females as a pre-emptive strategy to avoid this kind of reproductive failure.

Finally, a note of caution: I do not want to give the impression that females benefit unreservedly from polyandry. In sexually monogamous species, promiscuity either confers no benefits or imposes a cost on females. Even in species where polyandry is routine, female behaviour may be a trade-off between the costs and benefits. The most apparent cost of polyandry, at least for humans, is the risk of contracting a sexually transmitted disease, such as HIV, which may compromise the lifetime reproductive success of both females and males. The selection pressures created by HIV on humans are far too recent for us to have adapted, but intriguingly our more promiscuous cousin, the chimpanzee, seem to have done just that. Chimpanzees are relatively immune to HIV, but a human catching HIV from chimpanzees is defenceless.[56]

Conclusion

In the previous chapters I have tried to address the double-edged question of why sex is a battle. We have explored this issue principally by considering historical and biological evidence and, while not ignoring it, I have avoided becoming too embroiled in the politics of gender bias among researchers. There has undoubtedly been a battle between female and male researchers to force biologists to consider the female perspective, and many with an interest in feminism may feel that, while they may not have won, there has been considerable progress. From what was a distinctly androcentric starting point, recent research has made it abundantly clear that females are not passive participants in sexual reproduction. None the less, I think we also have to accept the truth of Geoff Parker's assertion, made over twenty years ago, that sexual selection acts more intensively on males than on females. That sexual selection pushes males harder than females is largely down to the size of their sex cells. When one set of contestants comprises a swarm of tiny, highly mobile sperm and the other a few large, immotile eggs, there is a certain inevitability about the type of battle that will ensue. The most important consequence of the difference between male and female sex cells is that the reproductive potential of males far outstrips that of females. The most successful males produce many more offspring than the most successful females. However, males do not have everything their own way and for species with internal fertilization, at least, the evolutionary battleground on which sexual conflict occurs – the female reproductive tract – is one designed by evolution to counter the ability of sperm to run circles round eggs. This is, perhaps, the most significant discovery of the past two decades – that male and female reproductive attributes co-evolve.[57] Adaptation and counter-adaptation – both sexes in a state of dynamic flux – each evolving, now, in response to adaptations in the other. Where males coerce females into copulating, females have a range of subtle behavioural and physiological counter-strategies. Where females evolve long and convoluted reproductive tracts better to regulate sperm uptake, males respond by evolving longer sperm. The idea that there exists a battle between the sexes implies that there are winners and losers, but if we think about sexual interactions as part of a co-evolutionary process, it is not obvious that either sex can ever be a clear winner. At any moment in time one sex may have slightly more control than the other, but the battle between the sexes is an evolutionary see-saw – subtle, sophisticated and inevitable.

References

Preface

1 Brenan (1978).
2 Clarke (1998), 240.
3 Laumann *et al.* (1994); Wellings *et al.* (1994).

1 Competition, Choice and Sexual Conflict

1 Massa (1981).
2 Jöchle (1971).
3 Jöchle (1984).
4 P. B. Medawar and J. S. Medawar (1984).
5 Aristotle (1943).
6 Dunbar (1995).
7 C. Darwin (1871); E. Darwin (1794, 1796); Colp (1986), King-Hele (1999).
8 Ghiselin (1969).
9 Trivers (1972).
10 Bateman (1948).
11 Parker (2000); Winge (1937); Parker, 'Sperm competition and its evolutionary consequences in the insects' (1970).
12 Williams (1966); Maynard Smith (1958); E. O. Wilson (1975); Dawkins (1976).
13 Horne (1988).
14 Parker (1984; 1998); Ruse (1999).
15 R. L. Smith, Preface (1984); Spallazani (1769); Birkhead (1997).
16 Darwin (1851); Burkhardt and Smith (1988).
17 Darwin (1875); Secord (1981).
18 Smith (1998).
19 Krause (1879); Raverat (1952).
20 Birkhead (1997).
21 Heape (1913).
22 Parker (1984); Trivers (1972).
23 Gowaty, *Feminism and Evolutionary Biology* (1996).
24 Kuhn (1970).
25 Woodward and Goodstein (1996).
26 Barkow *et al.* (1992); Buss (1994); Wright (1994). For an example of

different views on one aspect of evolutionary psychology, that of status and male reproductive success, see the 'Commentary' in *Behavioral & Brain Sciences* (1993), 16, 305–22.

27 Morell (1995).
28 R. L. Smith (1984b).
29 Baker and Bellis (1995); Kinsey *et al.* (1948; 1953); Masters and Johnson (1966); Robinson (1976).
30 Baker (1978; 1987). Critics of Baker's work on human navigation are cited in Baker (1987).
31 Baker and Bellis (1995).
32 Diamond (1986).
33 Cohen (1967); Baker and Bellis (1988).
34 Baker and Bellis (1995); Moore *et al.* (1999).
35 Birkhead, Moore and Bedford (1997).
36 Short (1997); Seuanez *et al.* (1977).
37 P. Robinson (1976).
38 Holland and Rice (1998).

2 Paternity and Protection

1 D. Robinson (1985).
2 Hrdy (1981).
3 Brenan (1957).
4 Aristotle (1943).
5 Nabours (1927).
6 Omura (1939).
7 Parker, 'Sperm competition and its evolutionary effect on copula duration...' (1970); Boorman and Parker (1976).
8 Birdsall and Nash (1973); Nelson and Hedgecock (1977).
9 Westneat (1987)
10 Jeffreys *et al.* (1985).
11 Burke and Bruford (1987); Wetton *et al.* (1987); see also Burke (1989).
12 Morton *et al.* (1990); Wagner *et al.* (1996).
13 Birkhead, 'Sperm competition in birds ...' (1998).
14 Hunter *et al.* (1992).
15 Westneat and Webster (1994).
16 Møller (1994).
17 Møller (1998).
18 Møller and Birkhead (1994).
19 R. L. Smith (1998).
20 Thornhill and Alcock (1983).
21 Ibid.
22 Regan (1925).
23 Pietsch (1976).
24 Kohda (1995).
25 A. G. Jones and Avise (1997); A. G. Jones *et al.* (1998).

26 Birkhead and Møller (1992).
27 *As You Like It*, I.iii.73–4.
28 Ribble (1991).
29 Kummer (1995).
30 Taggart *et al.* (1998).
31 White (1789).
32 Selous (1901).
33 M. D. Beecher and I. M. Beecher (1979).
34 Westneat (1994).
35 Parker (1970c).
36 Birkhead and Pringle (1986); Jormalainen (1998).
37 Gomendio *et al.* (1998).
38 Smith (1979).
39 Birkhead and Møller (1992).

3 Genitalia

1 Trivers (1972); Parker, 'Sperm competition and its evolutionary consequences in the insects' (1970).
2 Lind (1963).
3 Gilbert (1979).
4 T. Pissari (personal communication).
5 Lind (1963).
6 Van Drimmelen (1946).
7 Tauber (1875).
8 Bobr *et al.* (1964).
9 Birkhead and Møller (1992).
10 'The Cunning Little Tailor'.
11 Overstreet and Mahi-Brown (1993); Birkhead, Møller and Sutherland (1993).
12 Bakst *et al.* (1994); Overstreet and Mahi-Brown (1993).
13 Cohen and Tyler (1980).
14 R. E. Jones (1997).
15 Cohen (1998).
16 Howarth (1974); Lind (1963).
17 Gomendio *et al.* (1998).
18 Birkhead and Møller, 'Sexual selection ...' (1993); Racey (1979); Mayer (1995).
19 Birkhead and Møller, 'Sexual selection ...' (1993).
20 Ibid.
21 Schuett *et al.* (1997).
22 Pinto-Correia (1997).
23 Aristotle (1943).
24 Moore (1993); Pinto-Correia (1997).
25 R. E. Jones (1997).
26 Cohen (1999).

27 Van Voorhies (1992); but see Gems and Riddle (1996).
28 Olsson, Madsen and Shine (1997).
29 Afzelius (1995); Short (1977; 1979).
30 Birkhead and Møller (1992).
31 Short (1979).
32 Stockley *et al.* (1997).
33 Møller and Briskie (1995).
34 Gomendio *et al.* (1998).
35 Harcourt *et al.* (1995).
36 Dixson (1998).
37 Gagneux *et al.* (1997).
38 Robbins (1999).
39 Kinsey *et al.* (1948; 1953); Baker and Bellis (1995); Wellings *et al.* (1994)
40 Dixson (1998).
41 Amann (1981); Dixson (1998); Bedford (1994).
42 Aristotle (1943); Bedford (1977); Carrick and Setchell (1977); Werdelin and Nilsonne (1998).
43 Carrick and Setchell (1977); Freeman (1990).
44 Chance (1996).
45 Freeman (1990).
46 Bedford (1977).
47 Werdelin and Nilsonne (1998).
48 Riley (1938).
49 Wolfson (1954).
50 Fatio (1864); Birkhead, Briskie and Møller (1993).
51 Mann and Lutwak-Mann (1981).
52 Dixson (1998).
53 Martan and Shepherd (1976).
54 Dickinson and Rutowski (1989).
55 R. Mason (personal communication).
56 Koene and Chase (1998).
57 Hanlon and Messenger (1996).
58 Winterbottom *et al.* (1999).
59 Tokarz and Slowinski (1990).
60 Baur (1998).
61 Mann and Lutwak-Mann (1981).
62 Waage (1979).
63 R. Mason (personal communication).
64 Eberhard (1985).
65 Dixson (1987); Verrell (1992).
66 Arnqvist (1998).

4 Sperm, Ejaculates and Ova

1 Jöchle (1971).
2 Ibid.

3 Lind (1963).
4 Pinto-Correia (1997).
5 J. A. Moore (1993); Pinto-Correia (1997).
6 Pinto-Correia (1997).
7 Pinto-Correia (1997).
8 Laurila and Seppa (1998).
9 Ibid.
10 J. A. Moore (1993).
11 Ibid.
12 Pinto-Correia (1997).
13 J. A. Moore (1993).
14 Parker *et al.* (1972); Hurst (1990).
15 R. E. Jones (1997).
16 Weir (1971).
17 Afzelius (1995).
18 Bedford *et al.* (1984).
19 Pitnick, Markow and Spicer (1995).
20 Cummins and Woodall (1985); Gage (1997).
21 Roldan *et al.* (1992).
22 Briskie *et al.* (1997); Radwan (1996); LaMunyon and Ward (1998); Dixson (1998); Sivinski (1984).
23 Karr and Pitnick (1996).
24 Birkhead, Møller and Sutherland (1993).
25 LaMunyon and Ward (1998).
26 Briskie and Montgomerie (1993).
27 Briskie *et al.* (1997).
28 Beatty (1970); Wooley (1971).
29 Snook and Karr (1998).
30 He *et al.* (1995); Cook and Wedell (1999).
31 Parker (1984).
32 Cohen (1967; 1969).
33 Wallace (1974); Manning and Chamberlain (1994).
34 Jivoff (1997).
35 Petersen *et al.* (1992); Warner *et al.* (1995).
36 Petersen and Warner (1998); Parker, 'Sperm competition games: raffles and roles' (1990) and 'Sperm competition games: sneaks and extra-pair copulations' (1990).
37 R. E. Jones (1997); Suarez (1987).
38 R. E. Jones (1997).
39 Moore and Akhnodi (1996); Holt *et al.* (1997).
40 Froman and Feltman (1998).
41 Birkhead, *et al.* (1999).
42 Mann and Lutwak-Mann (1981).
43 Mann (1964); Mann and Lutwak-Mann (1981).
44 Riemann *et al.* (1967).
45 Fowler and Partridge (1989); Wolfner (1997).

46 Prout and Bundgaard (1977); Chapman *et al.* (1995); Harshman and Prout (1994); Clark *et al.* (1995).
47 Rice (1996).
48 Holland and Rice (1999).

5 Copulation, Insemination and Fertilization

1 Bourke and Franks (1995); Packer and Pusey (1983).
2 Levitan (1998).
3 Buchsbaum (1938); Caspers (1984).
4 Petersen *et al.* (1998).
5 Chimmo (1868).
6 Parker, 'Sperm competition and evolutionary consequences ...' (1970); Parker (1984).
7 Eady (1995).
8 Birkhead, 'Sperm Competition in Birds' (1998).
9 Mann and Lutwak-Mann (1981); Overstreet and Mahi-Brown (1993).
10 Thornhill and Alcock (1983); Hinton (1964); A. Stutt (personal communication).
11 Thornhill and Alcock (1983); Hinton (1964).
12 Thornhill and Alcock (1983).
13 Michiels and Newman (1998).
14 Norman and Lu (1997).
15 Nakamizo (1987).
16 Castro *et al.* (1996).
17 Wilkinson and Birkhead (1995).
18 Thornhill and Alcock (1983).
19 C. Darwin (1839).
20 Wilkelski and Baurle (1996).
21 K. Wilson (personal communication).
22 Bourke and Franks (1995).
23 Ridley (1988).
24 Birkhead, Atkin and Møller (1987).
25 Davies (1992); see also Emlen *et al.* (1998).
26 Birkhead and Møller, 'Why do male birds stop copulating ...?' (1993).
27 LeBoeuf (1974).
28 Carter and Getz (1993).
29 de Waal (1995).
30 Hoffer and East (1995).
31 Thornhill and Thornhill (1983); R. L. Smith, 'Human sperm competition' (1984).
32 Hill and Hurtardo (1996).
33 Thornhill and Alcock (1983).
34 Olsson (1995).
35 Hatziolos and Caldwell (1983).
36 R. E. Jones (1997).

37 Wedekind (1994).
38 R. E. Jones (1997).

6 Mechanisms of Sperm Competition and Sperm Choice

1 Bateman (1948).
2 Simmons and Siva-Jothy (1998).
3 Diesel (1990).
4 R. L. Smith (1979).
5 Waage (1979).
6 Gowaty (1994).
7 Gack and Peschke (1994).
8 Parker, 'Sperm competition games: raffles and roles' (1990) and 'Sperm competition games: sneaks and extra-pair copulations' (1990).
9 Hosken (1999); Simmons, Parker and Stockley (1999).
10 Aristotle (1943).
11 Martin *et al.* (1974).
12 Compton *et al.* (1978).
13 Cheng *et al.* (1983).
14 Lessells and Birkhead (1990).
15 Wishart (1987).
16 Birkhead and Biggins (1998).
17 Pellatt and Birkhead (1994).
18 Birkhead, Fletcher, Pellatt and Staples (1995); Colegrave *et al.* (1995)
19 Dewsbury (1984).
20 Schwagmeyer and Foltz (1990); Schwagmeyer (1995).
21 Huck *et al.* (1989).
22 Ginsberg and Huck (1989).
23 Cole and Davies (1914).
24 Dziuk (1996).
25 Bechstein (1881).
26 Steele and Wishart (1992).
27 Markow (1997).
28 Markow (1982); see also Clark, Begun and Prout (1999) for more details of male–female interactions in sperm competition.
29 Wedekind, Chapuisat, Macas and Rulicke (1995).
30 Eberhard (1996).
31 Thornhill (1983).
32 Eberhard (1996).
33 Carré and Sardet (1984); Carré *et al.* (1992).
34 Birkhead, 'Cryptic female choice . . .' (1998).
35 P. I. Ward (1993; 1997); Otronen *et al.* (1997); Simmons, Stockley, Jackson and Parker (1996).
36 Madsen *et al.* (1992).
37 Wildt *et al.* (1987).
38 Parker (1992).

39 Capula and Luiselli (1994).
40 Olsson *et al.* (1994); Olsson *et al.* (1996).
41 Desmond and Moore (1991).
42 Wilson *et al.* (1997).
43 Stockley (1997); Cunningham and Cheng (1999).
44 Keller and Reeve (1995).
45 Delph and Havens (1998); Bishop *et al.* (1996).
46 Birkhead, 'Cryptic female choice ...' (1998).
47 Kempenaers *et al.* (1996); Hasselquist *et al.* (1996).

7 The Benefits of Polyandry

1 Lumpkin (1981); Gowaty, 'Battles of the sexes ...' (1996).
2 Trivers (1972).
3 Bateman (1948).
4 Trivers (1972); Arnold and Duvall (1994).
5 Parker (1984).
6 S. M. Smith (1988).
7 Kempenaers *et al.* (1992).
8 Halliday and Arnold (1987); Cheng and Siegel (1990).
9 Jequier (1985).
10 Bray *et al.* (1975).
11 Ridley (1988).
12 Parker (1984; 1998).
13 Pitnick and Markow (1994); Pitnick, Spicer and Markow (1995).
14 Birkhead and Møller (1992).
15 Wetton and Parkin (1991); Lifjeld (1994); Birkhead, Veiga and Fletcher (1995).
16 Wedekind (1994); Austin and Short (1982).
17 Gray, 'Female red-winged blackbirds accrue material benefits ...' (1997).
18 Birkhead and Fletcher (1995).
19 Matthews *et al.* (1997).
20 Lodge *et al.* (1971).
21 Hoogland (1998).
22 Ketterson *et al.* (1997).
23 Vahed (1998).
24 Michiels and Streng (1998).
25 Fabre (1897); Gould (1984).
26 Lawrence (1992).
27 Andrade (1996).
28 Mills (1994).
29 de Waal (1995).
30 Hunter and Davis (1998).
31 Crook (1994).
32 Freuchen (1936); Burch (1975); Chance (1990).
33 Darwin (1839); Faabourg *et al.* (1995).

34 Davies (1992); Hartley and Davies (1994).
35 Gray, 'Intraspecific variation in extra-pair behaviour . . .' (1997).
36 Dixson (1998).
37 Daly and Wilson (1987).
38 Hill and Hurtado (1996).
39 McKinney *et al.* (1983).
40 Evarts and Williams (1987).
41 Westneat, Sherman and Morton (1990).
42 Andersson (1994).
43 Williams (1975).
44 Trivers (1972); J. A. Zeh and D. W. Zeh (1996; 1997).
45 Fisher (1930); Andersson (1994).
46 Andersson (1994).
47 Kirkpatrick and Ryan (1991).
48 Hamilton and Zuk (1982).
49 Sheldon and Ellegren (1999).
50 Potts *et al.* (1991).
51 Penn and Potts (1998); Wedekind and Furi (1997).
52 Møller (1994).
53 Cockburn (1998).
54 J. A. Zeh and D. W. Zeh (1996; 1997).
55 Newcomer *et al.* (1999).
56 Gao *et al.* (1999).
57 Parker (1998); Holland and Rice (1998).

Bibliography

Afzelius, B., 'Gustaf Retzius and spermatology', *International Journal of Developmental Biology* (1995), 39, 675–85

Amann, R. P., 'A critical review of methods for evaluation of spermatogenesis from seminal characteristics', *Journal of Andrology* (1981), 2, 37–58

Andersson, M., *Sexual Selection* (Princeton: Princeton University Press, 1994)

Andrade, M. C. B., 'Sexual selection for male sacrifice in the Australian redback spider', *Science* (1996), 271, 70–2

Andry du Bois-Regard, N., *An Account of the Breeding of Worms in Human Bodies* (London: H. Rhodes & A. Bell, 1701)

Anon., *The Complete Masterpiece: Displaying the Secrets of Nature in the Generation of Man, Aristotle's Works Completed*, vol. 1 (London, 1741)

Aristotle, *Generation of Animals*, trans. A. L. Peck (London: Heinemann, 1943)

Arnold, S. J., and D. Duvall, 'Animal mating systems: a synthesis based on selection theory', *American Naturalist* (1994), 143, 317–48

Arnqvist, G., 'Comparative evidence for the evolution of genitalia by sexual selection', *Nature* (1998), 393, 784–6

Austin, C. R., and R. V. Short, *Reproduction in Mammals*, vol. 4 (Cambridge: Cambridge University Press, 1982)

Baker, R. R., *The Evolutionary Ecology of Animal Migration* (London: Hodder & Stoughton, 1978)

– 'Human navigation and magnetoreception: the Manchester experiments do replicate', *Animal Behaviour* (1987), 35, 691–704

Baker, R. R., and M. A. Bellis, '"Kamikaze" sperm in mammals?', *Animal Behaviour* (1988), 36, 936–9

– *Human Sperm Competition* (London: Chapman & Hall, 1995)

Bakst, M. R., G. J. Wishart, and J. P. Brillard, 'Oviducal sperm selection, transport, and storage in poultry', *Poultry Science Review* (1994), 5, 117–43

Barkow, J. H., L. Cosmides, and J. Tooby, *The Adapted Mind: Evolutionary Psychology and the Generation of Culture* (New York: Oxford University Press, 1992)

Bateman, A. J., 'Intra-sexual selection in *Drosophila*', *Heredity* (1948), 2, 349–68

Baur, B., 'Sperm Competition in Molluscs', in T. R. Birkhead and A. P. Møller (eds), *Sperm Competition and Sexual Selection* (San Diego: Academic Press, 1998), 255–305

Beatty, R. A., 'The genetics of the mammalian gamete', *Biological Reviews* (1980), 45, 73–119

Bechstein, J. M., *Natural History of Cage Birds* (London: Groombridge, 1881)

Bedford, J. M., 'Evolution of the scrotum: the epididymis as the prime mover', in J. H. Calaby and C. H. Tyndale-Biscoe (eds), *Reproduction and Evolution* (Canberra: Australian Academy of Sciences, 1977), 171–82

– 'The status and state of the human epididymis', *Human Reproduction Update* (1994), 9, 2187–99

Bedford, J. M., J. C. Rodger, and W. G. Breed, 'Why so many mammalian spermatozoa – a clue from marsupials?', *Proceedings of the Royal Society of London* (1984), B, 221, 221–33

Beecher, M. D., and I. M. Beecher, 'Sociobiology of bank swallows: reproductive strategy of the male', *Science* (Washington; 1979), 205, 1282–5

Birdsall, D. A., and D. Nash, 'Occurrence of successful multiple insemination of females in natural populations of deeer mice (*Peromyscus maniculatus*)', *Evolution* (1973), 17, 106–10

Birkhead, T. R., 'Darwin on sex', *Biologist* (1997), 44, 397–9

– 'Cryptic female choice: criteria for establishing female sperm choice', *Evolution* (1998), 52, 1212–18

– 'Sperm Competition in Birds: Mechanisms and Functions', in T. R. Birkhead and A. P. Møller (eds), *Sperm Competition and Sexual Selection* (San Diego: Academic Press, 1998), 579–622

Birkhead, T. R., L. Atkin, and A. P. Møller, 'Copulation behaviour of birds', *Behaviour* (1987), 101, 101–38

Birkhead, T. R., and J. D. Biggins, 'Sperm competition mechanisms in birds: models and data', *Behavioural Ecology* (1998), 9, 253–60

Birkhead, T. R., J. V. Briskie, and A. P. Møller, 'Male sperm reserves and copulation frequency in birds', *Behavioural Ecology and Sociobiology* (1993), 32, 85–93

Birkhead, T. R., and F. Fletcher, 'Male phenotype and ejaculate quality in the zebra finch *Taeniopygia guttata*', *Proceedings of the Royal Society of London* (1995), B, 262, 329–34

Birkhead, T. R., F. Fletcher, E. J. Pellatt, and A. Staples, 'Ejaculate quality and the success of extra-pair copulations in the zebra finch', *Nature* (1995), 377, 422–3

Birkhead, T. R., J. G. Martinez, T. Burke and D. P. Froman, 'Sperm mobility determines the outcome of sperm competition on the domestic fowl', *Proceedings of the Royal Society of London* (1999), B, 266, 1759–64

Birkhead, T. R., and A. P. Møller, *Sperm Competition in Birds: Evolutionary Causes and Consequences* (London: Academic Press, 1992)

– 'Sexual selection and the temporal separation of reproductive events: sperm storage data from reptiles, birds and mammals', *Biological Journal of the Linnean Society* (1993), 50, 295–311

– 'Why do male birds stop copulating while their partners are still fertile?', *Animal Behaviour* (1993), 45, 105–18

Birkhead, T. R., A. P. Møller, and W. J. Sutherland, 'Why do females make it so difficult for males to fertilize their eggs?', *Journal of Theoretical Biology* (1993), 161, 51–60

Birkhead, T. R., H. D. M. Moore, and J. M. Bedford, 'Sex, science and sensationalism', *Trends in Ecology and Evolution* (1997), 12, 121–2

Birkhead, T. R., and S. Pringle, 'Multiple mating and paternity in *Gammarus pulex*', *Animal Behaviour* (1986), 34, 611–13

Birkhead, T. R., J. P. Veiga, and F. Fletcher, 'Sperm competition and unhatched eggs in the house sparrow', *Journal of Avian Biology* (1995), 26, 343–5

Bishop, J. D. D., C. S. Jones, and L. R. Noble, 'Female control of paternity in the internally fertilizing compound ascidian *Diplosoma listerianum*. II. Investigation of male mating success using RAPD markers', *Proceedings of the Royal Society of London* (1996), B, 263, 401–7

Bobr, L. W., F. W. Lorenz, and F. X. Ogasawara, 'Distribution of spermatazoa in the oviduct and fertility in domestic birds. 1. Residence sites of spermatazoa in fowl oviducts', *Journal of Reproduction and Fertility* (1964), 8, 39–47

Boorman, E., and G. A. Parker, 'Sperm (ejaculate) competition in *Drosophila melanogaster*, and the reproductive value of females to males in relation to female age and mating status', *Ecological Entomology* (1976), 1, 145–55

Bourke, A. F. G., and N. Franks, *Social Evolution in Ants* (Princeton: Princeton University Press, 1995)

Bray, O. E., J. K. Kennelly, and J. L. Guarino, 'Fertility of eggs produced on territories of vasectomized red-winged blackbirds', *Wilson Bulletin* (1975), 87, 187–95

Brenan, G., *South from Granada* (London: Hamish Hamilton, 1957)

– *Thoughts in a Dry Season* (Cambridge: Cambridge University Press, 1978)

Briffault, R., *The Mothers* (London: Allen & Unwin, 1927)

Briskie, J. V., and R. Montgomerie, 'Patterns of sperm storage in relation to sperm competition in passerine birds', *Condor* (1993), 95, 442–54

Briskie, J. V., R. Montgomerie, and T. R. Birkhead, 'The evolution of sperm size in birds', *Evolution* (1997), 51, 937–45

Buchsbaum, R., *Animals without Backbones* (Harmondsworth: Penguin, 1938)

Burch, E. S. J., *Eskimo Kinsmen: Changing Family Relationships in Northwest Alaska*, American Ethnological Society Monograph No. 59 (New York: West Publishing Company, 1975)

Burke, T., 'DNA fingerprinting and other methods for the study of mating success', *Trends in Ecology and Evolution* (1989), 4, 139–44

Burke, T., and M. W. Bruford, 'DNA fingerprinting in birds', *Nature* (1987), 327, 149–52

Burkhardt, F. H., and S. Smith, *The Correspondence of Charles Darwin, vol. 4: 1847–1850* (Cambridge: Cambridge University Press, 1988)

Buss, D. M., *The Evolution of Desire* (New York, Basic Books, 1994)

Capula, M., and L. Luiselli, 'Can female adders multiply?', *Nature* (1994), 369, 528

Carré, D., C. Rouvière, and C. Sardet, '*In vitro* fertilisation in ctenophores: sperm entry, mitosis, and the establishment of bilateral symmetry in *Beroë ovata*', *Developmental Biology* (1991), 147, 381–91

Carré, D., and C. Sardet, 'Fertilization and early development in *Beroë ovata*', *Developmental Biology* (1984), 105, 188–95

Carrick, F. N., and B. P. Setchell, 'The evolution of the scrotum', in J. H. Calaby and C. H. Tyndale-Biscoe (eds), *Reproduction and Evolution* (Canberra: Australian Academy of Science, 1977)

Carter, C. S., and L. L. Getz, 'Monogamy and the prairie vole', *Scientific American* (1993), June 70–6

Caspers, H., 'Spawning periodicity and habitat of the palolo worm *Eunice viridis* (Polychaeta: Eunicidae) in the Samoan Islands', *Marine Biology* (1984), 79, 229–36

Castro, I., E. Minot, R. Fordham, and T. R. Birkhead, 'Polygynandry, face-to-face copulation and sperm competition in the hihi *Notiomystis cincta* (Aves: Meliphagidae)', *Ibis* (1996), 138, 765–71

Chance, M. R. A., 'Reason for externalisation of the testis in mammals', *Journal of Zoology* (London; 1996), 239, 691–5

Chance, N. A., *The Inupiat and Arctic Alaska* (Fort Worth: Holt, Reinhart & Winston, 1990)

Chapman, T., L. F. Liddle, J. M. Kalb, M. F. Wolfner, and L. Partridge, 'Cost of mating in *Drosophila melanogaster* females is mediated by male accessory gland products', *Nature* (1995), 373, 241–4

Cheng, K. M., J. T. Burns, and F. McKinney, 'Forced copulation in captive mallards. III. Sperm competition', *Auk* (1983), 100, 302–10

Cheng, K. M., and P. B. Siegel, 'Quantitative genetics of multiple mating', *Animal Behaviour* (1990), 40, 406–7

Chimmo, W., 'A visit to the North-East coast of Labrador, during the Autumn of 1867, by H.M.S. *Gannet*, Commander, W. Chimmo, R.N.', *Royal Geographical Society Journal* (1868), 38, 258–81

Clark, A. G., M. Aguade, T. Prout, L. G. Harshman, and C. H. Langley, 'Variation in sperm displacement and its association with accessory gland gland protein in *Drosophila melanogaster*', *Genetics* (1995), 139, 189–201

Clark, A. G., D. J. Begun, and T. Prout, 'Female x male interactions in Drosophila sperm competition', *Science* (1999), 283, 217–20

Clarke, A. E., *Disciplining Reproduction* (Berkeley: University of California Press, 1998)

Cockburn, A.,'The benefits of female choice . . .', in 7th International Behavioural Ecology Congress (Asilomar, California, 1998), Abstract 82

Cohen, J., 'Correlation between chiasma frequency and sperm redundancy', *Nature* (1967), 215, 862–3

– 'Why so many sperms? An essay on the arithmetic of reproduction', *Science Progress London* (1969), 57, 23–41

– 'Spermatozoa and antibodies', *Molecular Human Reproduction* (1998), 4, 313–17

– 'The evolution of the sexual arena', in T. D. Glover and C. Barratt (eds), *Male Fertility and Infertility* (Cambridge: Cambridge University Press, 1999)

Cohen, J., and K. R. Tyler, 'Sperm populations in the female genital tract of the rabbit', *Journal of Reproduction and Fertility* (1980), 60, 213–18

Cole, L. J., and G. L. Davies, 'The effect of alcohol on the male germ cell, studied by means of double mating', *Science* (1914), 39, 476–7

Colegrave, N., T. R Birkhead, and C. M. Lessells, 'Sperm precedence in zebra finches does not require special mechanisms of sperm competition', *Proceedings of the Royal Society of London* (1995), B, 259, 223–8

Colp, R., 'The relationship of Charles Darwin to the ideas of his grandfather, Dr Erasmus Darwin', *Biography* (1986), 9, 1–24

Compton, M. M., H. P. Van Krey, and P. B. Siegel, 'The filling and emptying of the uterovaginal sperm-host glands in the domestic hen', *Poultry Science* (1978), 57, 1696–1700

Cook, P. A., and N. Weddell, 'Non-fertile sperm delay female remating', *Nature* (1998), 397, 486

Crook, J. H., 'Explaining Tibetan polyandry: socio-cultural, demographic, and biological perspectives', in J. H. Crook and H. A. Osmaston (eds), *Himalayan Buddhist Villages* (New Delhi: Shri Jainendra Press, 1994), 735–86

Cummins, J. M., and P. F. Woodall, 'On mammalian sperm dimensions', *Journal of Reproduction and Fertility* (1985), 75, 153–75

Cunningham, E. J. A., and K. M. Cheng, 'Biases in sperm utilisation in the mallard: no evidence for selection by females based on sperm genotype', *Proceedings of the Royal Society of London* (1999), B, 266, 905–10

Daly, M., and M. Wilson, 'Child abuse and other risks of not living with both parents,' *Ethological Sociobiology* (1987), 6, 197–210

Darwin, C., *Journal of Researches into the Geology and Natural History of the various countries visited by H.M.S. Beagle (1832–1836)* (London: H. Colburn, 1839)

– *A Monograph on the sub-class Cirripedia*, 2 vols (London: Ray Society, 1851, 1854)

– *The Descent of Man, and Selection in Relation to Sex* (London: John Murray, 1871)

– *The Variation of Animals and Plants under Domestication*, 2 vols (London: John Murray, 1875)

Darwin, E., *Zoonomia, or The Laws of Organic Life*, 2 vols (London: 1794; 1976)

Davies, N. B., *Dunnock Behaviour and Social Evolution* (Oxford: Oxford University Press, 1992)

Dawkins, R., *The Selfish Gene* (Oxford: Oxford University Press, 1976)

de Waal, F. B. M., 'Bonobo sex and society', *Scientific American*, March 1995, 58–64

Delph, L. F., and K. Havens, 'Pollen competition in flowering plants', in T. R. Birkhead and A. P. Møller (eds), *Sperm Competition and Sexual Selection* (San Diego: Academic Press, 1998), 149–73

Desmond, A., and J. Moore, *Darwin* (London: Michael Joseph, 1991)

Dewsbury, D. A., 'Sperm competition in murid rodents', in R. L. Smith (ed.), *Sperm Competition and the Evolution of Animal Mating Systems* (Orlando: Academic Press, 1984), 547–71

Diamond, J. M., 'Variation in human testis size', *Nature* (1986), 320, 488–9

Dickinson, J. L., and R. L. Rutowski, 'The function of the mating plug in the chalcedon checkerspot butterfly', *Animal Behaviour* (1989), 38, 154–62

Diesel, R., 'Sperm competition and reproductive success in the decapod *Inachus phalangium* (Majidae): a male ghost spider crab that seals off rivals' sperm', *Journal of Zoology* (London; 1990), 220, 213–23

Dixson, A. F., 'Observations on the evolution of the genitalia and copulatory behaviour of male primates', *Journal of Zoology* (London; 1987), 213, 423–43

– *Primate Sexuality. Comparative Studies of the Prosiminians, Monkeys, Apes and Human Beings* (Oxford: Oxford University Press, 1998)

Dunbar, R., *The Trouble with Science* (London: Faber and Faber, 1995)

Dziuk, P. J., 'Factors that influence the proportion of offspring sired by a male following heterospermic insemination', *Animal Reproduction Science* (1996), 43, 65–88

Eady, P., 'Why do male *Callosobruchus maculatus* beetles inseminate so many sperm?', *Behavioural Ecology and Sociobiology* (1995), 36, 25–32

Eberhard, W. G., *Sexual Selection and Animal Genitalia* (Cambridge, Massachusetts, and London: Harvard University Press, 1985)

– *Female Control: Sexual Selection by Cryptic Female Choice* (Princeton: Princeton University Press, 1996)

Emlen, S. T., P. H. Wrege, and M. S. Webster, 'Cuckoldry as a cost of polyandry in the sex-role-reversed wattled jacana, *Jacana jacana*', *Proceedings of the Royal Society of London* (1998), B, 265, 2359–64

Evarts, S., and C. J. Williams, 'Multiple paternity in a wild population of mallards', *Auk* (1987), 104, 597–602

Faaborg, J., P. G. Parker, L. Delay, T. J. de Vries, J. C. Bernardz, S. Maria Paz, J. Naranjo, and T. A. Waite, 'Confirmation of co-operative polyandry in the Galapagos hawk (*Buteo galapagoensis*)', *Behavioural Ecology and Sociobiology* (1995), 36, 83–90

Fabre, J. H. *Souvenirs entomologiques*, trans. A. T. de Mattos (Paris: C. Delgrave, 1897)

Fatio, M. V., 'Note sur une particularité de l'appareil reproducteur male chez *Accentor alpinus*', *Revue et Magazin de Zoologie Pure et Appliquee* (1864), 27, 65–7

Fisher, R. A., *The Genetical Theory of Natural Selection* (Oxford: Clarendon Press, 1930)

Flowerdew, J. R., *Mammals: Their Reproductive Biology and Population Ecology* (London: Edward Arnold, 1987)

Fowler, K., and L. Partridge, 'A cost of mating in female fruitflies', *Nature* (1989), 338, 760–61

Freeman, S., 'The evolution of the scrotum: a new hypothesis', *Journal of Theoretical Biology* (1990), 145, 429–45

Freuchen, P., *Arctic Adventure* (London: Heinemann, 1936)

Froman, D. P. & Feltman, A. J., 'Sperm mobility: a quantitative trait in the domestic fowl', *Biology of Reproduction* (1998), 58, 379–84

Gack, C., and K. Peschke, 'Spermathecal morphology, sperm transfer and a novel mechanism of sperm displacement in the rove beetle, *Aleochara curtula* (Coleoptera, Staphylinidae)', *Zoomorphology* (1994), 114, 227–37

Gage, M. J. G., 'Mammalian sperm morphometry', *Proceedings of the Royal Society of London* (1997), B, 265, 97–103

Gagneux, P., D. S. Woodruff, and C. Boesch, 'Furtive mating in female chimpanzees' *Nature* (1997), 387, 368–9

Gao F., E. Bailes, D. L. Robertson, Y. Chen, C. M. Rodenburg, S. F. Michael, L. B. Cummins, L. O. Arthur, M. Peeters, G. M. Shaw, P. M. Sharp and B. H. Hahn, 'Origin of HIV-1 in the chimpanzee *Pan troglodytes*' *Nature* (1999), 397, 436–441

Gems, D., and D. L. Riddle, 'Longevity in *Caenorhabditis elengans* reduced by mating but not gamete production', *Nature* (1996), 379, 723–5

Ghiselin, M. T., *The Triumph of the Darwinian Method* (Chicago: University of Chicago Press, 1969)

Gilbert, A. B., 'Female genital organs', in A. S. King and J. McLelland (eds), *Form and Function in Birds* (New York: Academic Press, 1979), 237–360

Ginsberg, J. R., and U. W. Huck, 'Sperm competition in mammals', *Trends in Ecology and Evolution* (1989), 4, 74–9

Glover, T. D., and J. B. Sale, 'The reproductive system of the male rock hyrax (*Procavia* and *Heterohyrax*)', *Journal of Zoology* (London; 1968), 156, 351–62

Gomendio, M., A. H. Harcourt, and E. R. S. Roldan, 'Sperm Competition in Mammals', in T. R. Birkhead and A. P. Møller (eds), *Sperm Competition and Sexual Selection* (San Diego: Academic Press, 1998), 667–755

Gould, S. J., 'Only his wings remained', *Natural History* (1984), 9, 10–18

Gowaty, P. A., 'Architects of sperm competition', *Trends in Ecology and Evolution* (1994), 9, 160–62

– 'Battles of the sexes and origins of monogamy', in J. M. Black (ed.), *Partnerships in Birds* (Oxford: Oxford University Press, 1996), 21–52

– (ed.), *Feminism and Evolutionary Biology* (New York: Chapman & Hall, 1996)

Gray, E. M. 'Female red-winged blackbirds accrue material benefits from copulating with extra-pair males', *Animal Behaviour* (1997), 53, 625–39

– 'Intraspecific variation in extra-pair behavior of red-winged blackbirds (*Agelaius phoeniceus*)', *Ornithological Monographs*, 1997, 61–80

Halliday, T., and S. J. Arnold, 'Multiple mating by females: a perspective from quantitative genetics', *Animal Behaviour* (1987), 35, 939–41

Hamilton, W. D., and M. Zuk, 'Heritable true fitness and bright birds: a role for parasites?', *Science* (1982), 218, 384–7

Hanlon, R. T., and J. B. Messenger, *Cephalopod Behaviour* (Cambridge: Cambridge University Press, 1996)

Harcourt, A. H., A. Purvis, and L. Liles, 'Sperm competition: mating system, not breeding season, affects testes size of primates', *Functional Ecology* (1995), 9, 468–76

Harshman, L. G., and T. Prout, 'Sperm displacement without sperm transfer in *Drosophila melanogaster*', *Evolution* (1994), 48, 758–66

Hartley, I. R., and N. B. Davies, 'Limits to cooperative polyandry in birds', *Proceedings of the Royal Society of London* (1994), B, 257, 67–73

Hasselquist, D., S. Bensch, and T. von Schantz, 'Correlation between male song repertoire, extra-pair paternity and offspring survival in the great reed warbler', *Nature* (1996), 381, 229–32

Hatziolos, M. E., and R. L. Caldwell, 'Role reversal in courtship in the stomatopod *Pseudosquilla ciliata* (Crustacea)', *Animal Behaviour* (1983), 31, 1077–87

He, Y., T. Tanaka, and T. Miyata, 'Eupyrene and apyrene sperm and their numerical fluctuations inside the female reproductive tract of the armyworm, *Pseudaletia separata*', *Journal of Insect Physiology* (1995), 41, 689–94

Heape, W., *Sex Antagonism* (London: Constable, 1913)

Hill, K., and A. M. Hurtado, *Ache Life History* (New York: Aldine de Gruyter, 1996)

Hinton, H. E., 'Sperm transfer and the evolution of haemocoelic insemination', in K. C. Highnam (ed.), *Insect Reproduction* (Royal Entomological Society of London Symposium 2, 1964), 95–107

Hoffer, H., and M. L. East, 'Virilized sexual genitalia as adaptations of spotted hyenas', *Revue Suisse de Zoologie* (1995), 102, 895–906.

Holland, B. and W. R. Rice, 'Chase-away sexual selection: Antagonistic seduction versus resistance', *Evolution* (1998), 52, 1–7

– 'Experimental removal of sexual selection reverses intersexual antagonistic coevolution and removes reproductive load', *Proceedings of the National Academy of Science* (USA; 1999), 96, 5083–8

Holt, C., W. V. Holt, H. D. M. Moore, H. C. B. Reed, and R. M. Curnock, 'Objectively measured boar sperm motility parameters correlate with the outcomes of on-farm inseminations. Results from two fertility trials', *Journal of Andrology* (1997), 18, 312–23

Hoogland, J., 'Why do female Gunnison prairie dogs copulate with more than one male?', *Animal Behaviour* (1998), 55, 351–9

Horne, A., *Macmillan 1894–1956* (London: Macmillan, 1988)

Hosken, D. J., 'Sperm displacement in yellow dung flies: a role for females', *Trends in Ecology and Evolution* (1999), 14, 251–2

Howarth, B., 'Sperm storage as a function of the female reproductive tract', in A. D. Johnson and C. E. Foley (eds), *The Oviduct and its Functions* (New York: Academic Press, 1974), 237–70

Hrdy, S. B., *The Woman That Never Evolved* (Cambridge, Massachusetts: Harvard University Press, 1981)

Huck, U. W., B. A. Tonias, and R. D. Lisk, 'The effectiveness of competitive male inseminations in golden hamsters, *Mesocricetus auratus*, depends on an interaction of mating order, time delay between males and the time of mating relative to ovulation', *Animal Behaviour* (1989), 37, 674–80

Hunter, F. M., T. A. Burke, and S. E. Watts, 'Frequent copulation as a method of paternity assurance in the Northern Fulmar', *Animal Behaviour* (1992), 44, 149–56

Hunter, F. M., and L. S. Davis, 'Female Adelie penguins acquire nest material from extrapair males after engaging in extrapair copulations', *Auk* (1998), 115, 526–8

Hurst, L., 'Parasite diversity and the evolution of diploidy, multicellularity and anisogamy', *Journal of Theoretical Biology* (1990), 144, 429–43

Jeffreys, A. J., V. Wilson, and S. L. Thein, 'Hypervariable "minisatellite" regions in human DNA', *Nature* (1985), 314, 67–73

Jequier, A. M., 'Non-therapy related pregnancies in the consorts of a group of men with obstructive azoospermia', *Andrologia* (1985), 17, 6–8

Jivoff, P., 'Sexual competition among male blue crab, *Callinectes sapidus*', *Biological Bulletin* (1997), 193, 368–80

Jöchle, W., 'Biology and pathology of reproduction in Greek mythology', *Contraception* (1971), 4, 1–13

– 'Traces of embryo transfer and artificial insemination in antiquity and the medieval age', *Theriogenology* (1984), 21, 80–3

Jones, A. G., and J. C Avise, 'Polygynandry in the dusky pipefish *Syngnathus floridae* revealed by microsatellite DNA marking', *Evolution* (1997), 51, 1611–22

Jones, A. G., C. Kvarnemo, G. I. Moore, L. Simmons, and J. C. Avise, 'Microsatellite evidence for genetic monogamy and sex-biased recombination in the Western Australian seahorse *Hippocampus angustus*', *Molecular Ecology* (1998), 7, 1497–505

Jones, R. E., *Human Reproductive Biology*, 2nd edn (San Diego: Academic Press, 1997)

Jormalainen, V., 'Precopulatory mate guarding in crustaceans: male competitive strategy and intersexual conflict', *Quarterly Review of Biology* (1998), 73, 275–304

Karr, T. L., and S. Pitnick, 'The ins and outs of fertilization', *Nature* (1996), 379, 405–6

Keller, L., and H. K. Reeve, 'Why do females mate with multiple males? The sexually selected sperm hypothesis', *Advances in the Study of Behaviour* (1995), 24, 291–315

Kempenaers, B., F. Andriaesen, V. A. J. Noordwijk, and A. A. Dhondt, 'Genetic similarity, inbreeding and hatching failure in blue tits: are unhatched eggs infertile?', *Proceedings of the Royal Society of London* (1996), B, 263, 179–85

Kempenaers, B., G. R. Verheyen, M. T. B. Van den Broeck, C. Van Broeckhoven, and A. A. Dhondt, 'Extra-pair paternity results from female preference for high-quality males in the blue tit', *Nature* (1992), 357, 494–6

Ketterson, E. D., P. G. Parker, S. A. Raouf, V. Nolan, C. Ziegenfus, and R. Chandler, 'The relative impact of extra-pair fertilizations on variation in male and female reproductive success in dark-eyed juncos (*Junco hyemalis*)', *Ornithological Monographs*, 1997, 81–101

King-Hele, D., *Erasmus Darwin* (London: de la Mare, 1999)

Kinsey, A. C., W. B. Pomeroy, and C. E. Martin, *Sexual Behavior in the Human Male* (Philadelphia: W. B. Saunders, 1948)

– *Sexual Behavior in the Human Female* (Philadelphia: W. B. Saunders, 1953)

Kirkpatrick, M., and M. J. Ryan, 'The evolution of mating preferences and the paradox of the lek', *Nature* (1991), 350, 33–8

Koene, J. M., and R. Chase, 'Changes in the reproductive system of the snail

Helix aspersa caused by mucus from the love dart', *Journal of Experimental Biology* (1998), 201, 2313–19

Kohda, M., M. Tanimura, M. Kikue-Nakamura, and S. Yamagishi, 'Sperm drinking by female catfishes: a novel mode of insemination', *Environmental Biology of Fish* (1995), 42, 1–6

Krause, E., *Erasmus Darwin ... with a Preliminary Notice by Charles Darwin* (London: John Murray, 1879)

Kuhn, T. S., *The Structure of Scientific Revolutions*, 2nd edn (Chicago: University of Chicago Press, 1970)

Kummer, H., *In Quest of the Sacred Baboon* (Princeton: Princeton University Press, 1995)

LaMunyon, C. W., and S. Ward, 'Larger sperm outcompete smaller sperm in the nematode *Caenorhabditis elegans*', *Proceedings of the Royal Society of London* (1998), B., 265, 1997–2000

Laumann, E. P., J. H. Gagnon, R. T. Michael, and S. Michaels, *The Social Organization of Sexual Practices in the United States* (Chicago: University of Chicago Press, 1994)

Laurila, A., and P. Seppa, 'Multiple paternity in the common frog (*Rana temporaria*): genetic evidence from tadpole kin groups', *Biological Journal of the Linnean Society* (1998), 63, 221–32

Lawrence, S. E., 'Sexual cannibalism in the praying mantid *Mantis reliogosa*', *Animal Behaviour* (1992), 43, 569–83

LeBoeuf, B. J., 'Male–male competition and reproductive success in elephant seals', *American Zoologist* (1974), 14, 163–76

Lessells, C. M., and T. R. Birkhead, 'Mechanisms of sperm competition in birds: mathematical models', *Behavioural Ecology and Sociobiology* (1990), 27, 325–37

Levitan, D. R., 'Sperm limitation, gamete competition, and sexual selection in external fertilizers', in T. R. Birkhead and A. P. Møller (eds), *Sperm Competition and Sexual Selection* (San Diego: Academic Press, 1998), 175–217

Lifjeld, J. T., 'Do female house sparrows copulate with extra-pair mates to enhance their fertility?', *Journal of Avian Biology* (1994), 25, 75–6

Lind, L. R., *Aldrovandi on Chickens* (Norman: University of Oklahoma Press, 1963)

Lodge, J. R., N. S. Fechheimer, and R. G. Jaap, 'The relationship of *in vivo* sperm storage interval to fertility and embryonic survival in the chicken', *Biology of Reproduction* (1971), 5, 252–7

Lumpkin, S., 'Avoidance of cuckoldy in birds: the role of the female', *Animal Behaviour* (1981), 29, 303–4

McKinney, F., S. R. Derrickson, and P. Mineau, 'Forced copulation in waterfowl', *Behaviour* (1983), 86, 250–94

Madsen, T., R. Shine, J. Loman, and T. Hakansson, 'Why do female adders copulate so frequently?', *Nature* (1992), 355, 440–1

Mann, T., *The Biochemistry of Semen and of the Male Reproductive Tract* (London: Methuen, 1964)

Mann, T., and C. Lutwak-Mann, *Male Reproductive Function and Semen* (Berlin: Springer, 1981)
Manning, J. T., and A. T. Chamberlain, 'Sib-competition and sperm competiveness: an answer to "Why so many sperms?" and the recombination/sperm number correlation', *Proceedings of the Royal Society of London* (1994), B, 256, 177–82
Markow, T. A., 'Mating systems of cactophilic Drosophila', in J. S. F. Barker and W. T. Starmer (eds), *Ecological Genetics and Evolution: the Cactus–Yeast–Drosophila Model System* (New York: Academic Press, 1982), 273–87
– 'Assortative fertilization in Drosophila', *Proceedings of the National Academy of Sciences* (USA; 1997), 94, 7756–60
Martan, J., and B. A. Shepherd, 'The role of the copulatory plug in reproduction of the guinea pig', *Journal of Experimental Zoology* (1976), 196, 79–84
Martin, P. A., T. J. Reimers, J. R. Lodge, and P. J. Dziuk, 'The effect of ratios and numbers of spermatozoa mixed from two males on proportions of offspring', *Journal of Reproduction and Fertility* (1974), 39, 251–8
Massa, R., 'Muore un falco: l'onore e salvo' ['The hawk dies the honour is saved'], *Airone* (1981), 7, 21–5
Masters, W. H., and V. E. Johnson, *Human Sexual Response* (London: J. & A. Churchill, 1966)
Matthews, I. M., J. P. Evans, and A. E. Magurran, 'Male display rate reveals ejaculate characteristics in the Trinidadian guppy *Poecilia reticulata*', *Proceedings of the Royal Society of London* (1997), B, 264, 695–700
Maxwell, K., *A Sexual Odyssey* (New York: Plenum, 1996)
Mayer, F., 'Multiple paternity and sperm competition in the noctule bat (*Nyctalus noctule*) revealed by DNA fingerprinting', *Bat Research News* (1995), 36, 88
Maynard Smith, J., *The Theory of Evolution* (Harmondsworth: Penguin Books, 1958)
Medawar, P. B., and J. S. Medawar, *Aristotle to Zoos: A Philosophical Dictionary of Biology* (London: Weidenfeld & Nicolson, 1984)
Michiels, N. K., and L. J. Newman, 'Sex and violence in hermaphrodites', *Nature* (1998), 391, 647
Michiels, N. K., and A. I. Streng, 'Sperm exchange in a simultaneous hermaphrodite', *Behavioural Ecology and Sociobiology* (1998), 42, 171–8
Mills, J. A., 'Extra-pair copulations in the red-billed gull: females with high quality, attentive males resist', *Behaviour* (1994), 128, 41–64
Møller, A. P., *Sexual Selection and the Barn Swallow* (Oxford: Oxford University Press, 1994)
– 'Sperm competition and sexual selection', in T. R. Birkhead and A. P. Møller (eds), *Sperm Competition and Sexual Selection* (San Diego: Academic Press, 1998), 55–90
Møller, A. P., and T. R. Birkhead, 'The evolution of plumage brightness in birds is related to extra-pair paternity', *Evolution* (1994), 48, 1089–100
Møller, A. P., and J. V. Briskie, 'Extra-pair paternity, sperm competition and

the evolution of testis size in birds', *Behavioural Ecology and Sociobiology* (1995), 36, 357–65

Moore, H. D. M., and M. A. Akhondi, 'Fertilizing capacity of rat spermatozoa is correlated with decline in straight-line velocity measured by continuous computer-aided sperm analysis: epididymal rat spermatozoa from the proximal cauda have a greater fertilizing capacity *in vitro* than those from the distal cauda or vas deferens', *Journal of Andrology* (1996), 17, 50–60

Moore, H. D. M., M. Martin and T. R. Birkhead, 'No evidence for killer sperm or other selective interactions between human spermatozoa in ejaculates of different males *in vitro*', *Proceedings of the Royal Society of London* (1999), B, 266, 2343–2350

Moore, J. A., *Science as a Way of Knowing* (Cambridge, Massachusetts: Harvard University Press, 1993)

Morell, V., *Ancestral Passions* (New York: Simon & Schuster, 1995)

Morris, D., *The Naked Ape* (London: Jonathan Cape, 1967)

Morton, E. S., L. Forman, and M. Braun, 'Extrapair fertilizations and the evolution of colonial breeding in purple martins', *Auk* (1990), 107, 275–83

Nabours, R. K., 'Polyandry in the grouse locust, *Paratettix texanus* Hancock, with notes on inheritance of acquired characters and telegony', *American Naturalist* (1927), 61, 531–8

Nakamizo, M., 'Foriegn body of the oral cavity and the oropharynx sperm sacs of squid', *Otolaryngology* (1987), 59, 245–8

Nelson, K., and D. Hedgecock, 'Electrophoretic evidence of multiple paternity in the lobster *Homarus americanus* (Milne-Edwards)', *American Naturalist* (1977), 111, 361–5

Newcomer, S. D., Zeh, J. A. and Zeh, D. W.,'Genetic benefits enhance the reproductive success of polyandrous females.' *Proceedings of the National Academy of Sciences, USA* (1999), 96, 10236–41

Norman, M. D., and C. C. Lu, 'Sex in giant squid', *Nature* (1997), 389, 683–4

Olsson, M., 'Forced copulation and costly female resistance behavior in the Lake Eyre Dragon *Ctenophorus maculosus*', *Herpetologica* (1995), 51, 19–24

Olsson, M., A. Gullberg, H. Tegelstrom, T. Madsen, and R. Shine, 'Can female adders multiply?', *Nature* (1994), 369, 528

Olsson, M., T. Madsen, and R. Shine, 'Is sperm really so cheap? Costs of reproduction in male adders, *Vipera berus*', *Proceedings of the Royal Society of London* (1997), B, 264, 455–9

Olsson, M., R. Shine, T. Madsen, A. Gullberg, and H. Tegelstrom, 'Sperm selection by females', *Nature* (1996), 383, 585

Omura, S., 'Selective fertilisation in *Bombyx mori*', *Japanese Journal of Genetics* (1939), 15, 29–35 (English résumé)

Otronen, M., P. Reguera, and P. I. Ward, 'Sperm storage in the yellow dungfly *Scathophaga stercoraria*: identifying the sperm of competing males in separate female spermathecae', *Ethology* (1997), 103, 844–54

Overstreet, J. W., and C. A. Mahi-Brown, 'Sperm processing in the female reproductive tract', in P. D. Griffin and P. M. Johnson (eds), *Local*

Immunity in Reproduction Tract Tissues (Oxford: Oxford University Press, 1993), 321–34

Packer, C., and A. E. Pusey, 'Adaptations of female lions to infanticide by incoming males', *American Naturalist* (1983), 121, 716–28

Parker, G. A., 'Sperm competition and its evolutionary consequences in the insects', *Biological Reviews* (1970), 45, 525–67

– 'Sperm competition and its evolutionary effect on copula duration in the fly *Scatophaga stercoraria*', *Journal of Insect Physiology* (1970), 16, 1301–28

– 'The reproductive behaviour and the nature of sexual selection in *Scatophaga stercoraria* L. (Diptera: Scatophagidae). VII. The origin and evolution of the passive phase', *Evolution* (1970), 24, 774–88

– 'Sperm competition and the evolution of animal mating strategies', in R. L. Smith (ed.), *Sperm Competition and the Evolution of Animal Mating Systems* (Orlando: Academic Press, 1984), 1–60

– 'Sperm competition games: raffles and roles', *Proceedings of the Royal Society of London* (1990), B, 242, 120–6

– 'Sperm competition games: sneaks and extra-pair copulations', *Proceedings of the Royal Society of London*, (1990b) B, 242, 127–33

– 'Snakes and female sexuality', *Nature* (1992), 355, 395–6

– 'Sperm competition and the evolution of ejaculates: towards a theory base', in T. R. Birkhead and A. P. Møller (eds), *Sperm Competition and Sexual Selection* (San Diego: Academic Press, 1998), 3–54

– 'Golden flies, sunlit meadows: a tribute to the yellow dungfly', in L. A. Dugatkin (ed.), *Model Systems in Behavioral Ecology* (Princeton: Princeton University Press, 2000)

Parker, G. A., R. R. Baker, and V. G. F. Smith, 'The origin and evolution of gamete dimorphism and the male–male phenomenon', *Journal of Theoretical Biology* (1972), 36, 529–53

Pellatt, E. J., and T. R. Birkhead, 'Ejaculate size in zebra finches *Taeniopygia guttata* and a method for obtaining ejaculates from passerine birds', *Ibis* (1994), 136, 97–101

Penn, D., and W. K. Potts, 'Chemical signals and parasite-mediated sexual selection', *Trends in Ecology and Evolution* (1998), 13, 391–6

Petersen, C. W., and R. R. Warner, 'Sperm competition in fishes', in T. R. Birkhead and A. P. Møller (eds), *Sperm Competition and Sexual Selection* (San Diego: Academic Press, 1998), 435–63

Petersen, C. W., R. R. Warner, S. Cohen, H. C. Hess, and A. T. Sewell, 'Variable pelagic fertilization success: implications for mate choice and spatial patterns of mating', *Ecology* (1992), 73, 391–401

Pietsch, T. W., 'Dimorphism, parasitism and sex: reproductive strategies among deepsea ceratioid anglerfishes', *Coepia* (1976), 4, 781–93

Pinto-Correia, C., *The Ovary of Eve: Egg and Sperm and Preformation* (Chicago: University of Chicago Press, 1997)

Pitnick, S., and T. A. Markow, 'Male gametic strategies: sperm size, testes size, and the allocation of ejaculate among successive mates by the sperm-limited fly *Drosophila pachea* and its relatives', *American Naturalist* (1994), 143, 785–819

Pitnick, S., T. A. Markow, and G. S. Spicer, 'Delayed male maturity is a cost of producing large sperm in *Drosophila*', *Proceedings of the National Academy of Sciences* (USA; 1995), 92, 10,614–18

Pitnick, S., G. S. Spicer, and T. A. Markow, 'How long is a giant sperm?' *Nature* (1995), 375, 109

Potts, W. K., C. J. Manning, and E. K. Wakeland, 'Mating patterns in semi-natural populations of mice influenced by MHC genotype', *Nature* (1991), 352, 619–21

Prout, T., and J. Bundgaard, 'The population genetics of sperm displacement', *Genetics* (1977), 85, 95–121

Racey, P. A., 'The prolonged storage and survival of spermatazoa in Chiroptera', *Journal of Reproduction and Fertility* (1979), 56, 403–16

Radwan, J., 'Intraspecific variation in sperm competition success in the bulb mite: a role for sperm size', *Proceedings of the Royal Society of London* (1996), B, 263, 855–9

Raverat, G., *Period Piece: A Cambridge Childhood* (London: Faber and Faber, 1952)

Regan, T. C., 'Dwarfed males parasitic on the females in oceanic anglerfishes (Pediculati Ceratioidra)', *Proceedings of the Royal Society of London* (1925), B, 97, 386–400

Ribble, D. O. 'The monogamous mating system of *Peromyscus californicus* as revealed by DNA fingerprinting', *Behavioural Ecology and Sociobiology* (1991), 29, 161–6

Rice, W. R., 'Sexually antagonistic male adaptation triggered by experimental arrest of female evolution', *Nature* (1996), 381, 232–43

Ridley, M., 'Mating frequency and fecundity in insects', *Biological Reviews* (1988), 63, 509–49

Riemann, J. G., D. J. Moen, and B. J. Thorson, 'Female monogamy and its control in the house fly, *Musca domestica*', *Journal of Insect Physiology* (1967), 13, 407–18

Riley, G. M., 'Cytological studies on spermatogenesis in the house sparrow, *Passer domesticus*', *Cytologia* (1938), 9, 165–76

Robbins, M. M., 'Male mating patterns in wild multimale mountain gorilla groups', *Animal Behaviour* (1999), 57, 1013–20

Robinson, P., *The Modernization of Sex* (Ithaca, New York: Cornell University Press, 1976)

Robinson, D., *Chaplin: His Life and Art* (London: Collins, 1985)

Roldan, E. R. S., M. Gomendio, and A. D. Vitullo, 'The evolution of eutherian spermatozoa and underlying selective forces: female selection and sperm competition', *Biological Reviews* (1992), 67, 1–43

Ruse, M., *Mystery of Mysteries* (Cambridge, Massachusetts: Harvard University Press, 1999)

Schuett, G. W., P. J. Fernandez, W. F. Gergits, N. J. Casna, D. Chiszar, H. M. Smith, J. B. Mitton, S. P. Mackessy, R. A. Odum, and M. J. Demlong, 'Production of offspring in the absence of males: evidence for facultative parthenogenesis in bisexual snakes', *Herpetological Natural History* (1997), 5, 1–10

Schwagmeyer, P. L., 'Searching today for tomorrow's mates', *Animal Behaviour* (1995), 50, 759–67

Schwagmeyer, P. L., and D. W. Foltz, 'Factors affecting the outcome of sperm competition in 13-lined ground squirrels', *Animal Behaviour* (1990), 39, 156–62

Secord, J. A., 'Nature's Fancy: Charles Darwin and the Breeding of Pigeons', *Isis* (1981), 72, 163–86

Selous, E., *Bird Watching* (London: J. M. Dent, 1901)

Seuanez, H. N., A. D. Carothers, D. E. Martin, and R. V. Short, 'Morphological abnormalities in spermatozoa of man and great apes', *Nature* (1977), 270, 345–7

Sheldon, B. C., and H. Ellegren, 'Sexual selection resulting from extrapair paternity in collared flycatchers', *Animal Behaviour* (1999), 57, 285–98

Short, R. V., 'Sexual selection and the descent of man', in *Proceedings of the Canberra Symposium on Reproduction and Evolution. Australian Academy of Sciences* (1977), 3–19

– 'Sexual selection and its component parts, somatic and genital selection as illustrated by man and the great apes', *Advances in the Study of Behaviour* (1979), 9, 131–58

– review of R. R. Baker and M. A. Bellis, *Human Sperm Competition: Copulation, Masturbation and Infidelity, European Sociobiological Society* (1997), 47, 20–23

Simmons, L. W., G. A. Parker, and P. Stockley, 'Sperm displacement in the yellow dungfly, *Scatophaga stercoraria*: an investigation of male and female processes', *American Naturalist* (1999), 153, 302–14

Simmons, L. W., and M. T. Siva-Jothy, 'Sperm competition in insects: mechanisms and the potential for selection', in T. R. Birkhead and A. P. Møller (eds), *Sperm Competition and Sexual Selection* (London: Academic Press, 1998)

Simmons, L. W., P. Stockley, R. L. Jackson, and G. A. Parker, 'Sperm competition or sperm selection: no evidence for female influence over paternity in yellow dung flies *Scatophaga stercoraria*', *Behavioural Ecology and Sociobiology* (1996), 38, 199–206

Sivinski, J., 'Sperm in competition', in R. L. Smith (ed.), *Sperm Competition and the Evolution of Animal Mating Systems* (Orlando: Academic Press, 1984), 223–49

Smith, R. L., 'Repeated copulation and sperm precedence: paternity assurance for a male brooding water bug', *Science* (1979), 205, 1029–31

– Preface to R. L. Smith (ed.) *Sperm Competition and the Evolution of Animal Mating Systems* (San Diego: Academic Press, 1984)

– 'Human sperm competition', in R. L. Smith (ed.) *Sperm Competition and the Evolution of Animal Mating Systems* (Orlando: Academic Press, 1984), 601–59

– Foreword to T. R. Birkhead and A. P. Møller (eds), *Sperm Competition and Sexual Selection* (San Diego: Academic Press, 1998)

Smith, S. M., 'Extra-pair copulations in black-capped chickadees: the role of the female', *Behaviour* (1988), 107, 15–23

Snook, R. R., and T. L. Karr, 'Only long sperm are fertilisation-competent in six sperm-heteromorphic *Drosophila* species', *Current Biology* (1998), 8, 291–4

Spallazani, L., *An Essay on Animal Reproduction* (London, 1769)

Steele, M. G., and G. J. Wishart, 'Evidence for a species-specific barrier to sperm transport within the vagina of the chicken-hen', *Theriogenology* (1992), 38, 1107–14

Stevenson, I., 'Male-biased mortality in Soay sheep', unpublished Ph.D. thesis (1994), University of Cambridge

Stockley, P., 'No evidence of sperm selection by female common shrews', *Proceedings of the Royal Society of London* (1997), B., 264, 1497–500

Stockley, P., M. J. G. Gage, G. A. Parker, and A. P. Møller, 'Sperm competition in fishes: the evolution of testis size and ejaculate characteristics', *American Naturalist* (1997), 149, 933–54

Suarez, S. S., 'Sperm transport and motility in the mouse oviduct: observation in situ', *Biology of Reproduction* (1987), 36, 203–10

Taggart, D. A., W. G. Breed, P. D. Temple-Smith, A. Purvis, and G. Shimmin, 'Reproduction, mating strategies and sperm competition in marsupials and monotremes', in T. R. Birkhead and A. P. Møller (eds), *Sperm Competition and Sexual Selection* (San Diego: Academic Press, 1998), 623–66

Tauber, P., 'Om hønseæggets befrugtning i æggelederen', *Naturhistorisk Tidsskrift* (1875), 10, 63–106

Thornhill, R., 'Cryptic female choice and its implications in the scorpionfly *Harpobittacus nigriceps*', *American Naturalist* (1983), 122, 765–88

Thornhill, R., and J. Alcock, *The Evolution of Insect Mating Systems* (Cambridge, Massachusetts and London: Harvard University Press, 1983)

Thornhill, R., and N. M. Thornhill, 'Human rape: an evolutionary analysis', *Ethological Sociobiology* (1983), 4, 137–73

Tokarz, R. R., and J. B. Slowinski, 'Alternation of hemipenis use as a behavioural means of increasing sperm transfer in the lizard *Anolis sagrei*', *Animal Behaviour* (1990), 40, 374–9

Trivers, R. L., 'Parental investment and sexual selection', in B. Campbell (ed.), *Sexual Selection and the Descent of Man*, 1871–1971 (Chicago: Aldine-Atherton, 1972), 136–79

Vahed, K., 'The function of nuptial feeding in insects: a review of empirical studies', *Biological Reviews* (1998), 73, 43–78

Van Drimmelen, G. C., '"Sperm Nests" in the oviduct of the domestic hen', *Journal of the South African Veterinary Medical Association* (1946), 17, 42–52

Van Voorhies, W. A., 'Production of sperm reduces nematode lifespan', *Nature* (1992), 360, 456–8

Verrell, P. A., 'Primate penile morphologies and social systems: further evidence for an association', *Folia Primatologia* (1992), 59, 114–20

Waage, J. K., 'Dual function of the damselfly penis: sperm removal and transfer', *Science* (1979), 203, 916–18

Wagner, R. H., M. D. Schug, and E. S. Morton, 'Condition-dependent control of paternity by female purple martins: implications for coloniality',

Behavioural Ecology and Sociobiology (1996), 38, 379–89
Wallace, H., 'Chiasmata have no effect on fertility', *Heredity* (1974), 33, 423–9
Ward, P. I., 'Females influence sperm storage and use in the yellow dung fly *Scathophaga stercoraria* (L.)', *Behavioural Ecology and Sociobiology* (1993), 32, 313–19
– 'A possible explanation for cryptic female choice in the yellow dung fly *Scathophaga stercoraria* (L.)', *Ethology* (1997), 104, 97–110
Ward, S., and J. S. Carrel, 'Fertilization and sperm competition in the nematode *Caenorhabditis elegans*', *International Journal of Developmental Biology* (1979), 73, 304–21
Warner, R. R., D. Y. Shapiro, A. Marcanato, and C. W. Petersen, 'Sexual conflict: males with highest mating success convey the lowest fertilization benefits to females', *Proceedings of the Royal Society of London* (1995), B, 262, 135–9
Wedekind, C., 'Mate choice and maternal selection for specific parasite resistances before, during and after fertilisation', *Philosophical Transactions of the Royal Society of London* (1994), B, 346, 303–11
Wedekind, C., M. Chapuisat, E. Macas, and T. Rulicke, 'Non-random fertilization in mice correlates with MHC and something else', *Heredity* (1995), 77, 400–9
Wedekind, C., and S. Furi, 'Body odour preferences in men and women: do they aim for specific MHC combinations or simply heterozygosity?', *Proceedings of the Royal Society of London* (1997), B, 264, 1471–9
Weir, B. J., 'The reproductive organs of the female Plains Viscacha (*Lagostomus maximus*)', *Journal of Reproduction and Fertility* (1971), 25, 365–73
Wellings, K., J. Field, A. M. Johnson, and J. Wadsworth, *Sexual Behaviour in Britain* (Harmondsworth: Penguin Books, 1994)
Werdelin, L., and A. Nilsonne, 'The evolution of the scrotum and testicular descent in mammals: a phylogenetic view', *Journal of Theoretical Biology* (1998), 196, 61–72
Westneat, D. F., 'Extra-pair fertilizations in a predominantly monogamous bird: genetic evidence', *Animal Behaviour* (1987), 35, 877–86
– 'To guard or go forage: conflicting demands affect the paternity of male red-winged blackbirds', *American Naturalist* (1994), 144, 343–54
Westneat, D. F., P. W. Sherman, and M. L. Morton, 'The ecology and evolution of extra-pair copulations in birds', *Current Ornithology* (1990), 7, 331–69
Westneat, D. F., and M. S. Webster, 'Molecular analysis of kinship in birds: interesting questions and useful techniques', in B. Schierwater, B. Streit, G. P. Wagner and R. DeSalle (eds), *Molecular Ecology and Evolution: Approaches and Applications* (Basel: Birhauser Verlag, 1994), 91–126
Wetton, J. H., R. E. Carter, D. T. Parkin, and D. Walters, 'Demographic study of a wild house sparrow population by DNA fingerprinting', *Nature* (1987), 327, 147–9
Wetton, J. H., and D. T. Parkin, 'An association between fertility and

cuckoldry in the house sparrow *Passer domesticus*', *Proceedings of the Royal Society of London* (1991), B, 245, 227–33
White, G., *The Natural History and Antiquities of of Selborne* (London: 1789)
Wildt, D. E., M. Bush, K. L. Goodrowe, C. Packer, A. E. Pusey, J. L. Brown, P. Joslin, and S. J. O'Brian, 'Reproductive and genetic consequences of founding isolated lion populations', *Nature* (1987), 329, 328–31
Wilkelski, M., and S. Baurle, 'Pre-copulatory ejaculation solves time constraints during copulations in marine iguanas', *Proceedings of the Royal Society of London* (1996), B., 263, 439–44
Wilkinson, R., and T. R. Birkhead, 'Copulation behaviour in the vasa parrots *Coracopsis vasa* and *C. nigra*', *Ibis* (1995), 137, 117–19
Williams, G. C., *Adaptation and Natural Selection* (Princeton: Princeton University Press, 1966)
– *Sex and Evolution* (Princeton: Princeton University Press, 1975)
Wilson, E. O., *Sociobiology: The Modern Synthesis* (Cambridge, Massachusetts: Harvard University Press, 1975)
Wilson, N., S. Tubman, and P. Eady, 'Female genotype affects male success in sperm competition', *Proceedings of the Royal Society of London* (1997), B, 264, 1491–5
Winge, Ø., 'Succession of broods in *Lebistes*', *Nature* (1937), 140, 467
Winterbottom, M., T. R. Birkhead, and T. Burke, 'A phalloid organ and orgasm in a promiscuous bird', *Nature* (1999), 399, 28
Wishart, G. J., 'Regulation of the length of the fertile period in the domestic fowl by numbers of oviductal spermatozoa as reflected by those trapped in laid eggs', *Journal of Reproduction and Fertility* (1987), 80, 493–8
Wolfner, M. F., 'Tokens of love: functions and regulation of Drosophila male accessory gland products', *Insect Biochemistry and Molecular Biology* (1997), 27, 179–92
Wolfson, A., 'Sperm storage at lower-than-body temperature outside the body cavity in some passerine birds', *Science* (1954), 120, 68–71
Woodward, J., and D. Goodstein, 'Conduct, misconduct and the structure of science', *American Scientist* (1996), 84, 479–90
Wooley, D. M., 'Selection for the length of the spermatozoan midpiece in the mouse', *Genetical Research, Cambridge* (1971), 16, 261–75
Wright, R., *The Moral Animal* (London: Little, Brown, 1994)
Zeh, J. A., and D. W. Zeh, 'The evolution of polyandry I: intragenomic conflict and genetic incompatibility', *Proceedings of the Royal Society of London* (1996), B, 263, 1711–17
– 'The evolution of polyandry II: post-copulatory defences against genetic incompatibility', *Proceedings of the Royal Society of London* (1997), B, 264, 69–75

Appendix: Species mentioned in the text

Adder *Vipera berus*
Adelie penguin *Pygoscelis adeliae*
Alpine accentor *Prunella collaris*
Army worm *Pseudaletia separata*
Baboon, hamadryas *Papio hamadryas*
Bank swallow (see sand martin)
Bedbug *Cimex lectularius*
Beroë ovata
Bighorn sheep *Ovis canadensis*
Black-bellied fruitfly *Drosophila melanogaster*
Black-capped chickadee *Parus atricapillus*
Blackbird *Turdus merula*
Blue crab *Callinectes sapidus*
Blue tit *Parus caeruleus*
Blue-headed wrasse *Thalassoma bifasciatum*
Bonobo *Pan paniscus*
Buffalo weaver *Bubalornis niger*
Bulb mite *Rhizoglyphus robini*
Caenorhabditis elegans
California mouse *Peromyscus californicus*
Capelin *Mallotus villosus*
Chaffinch *Fringilla coelebs*
Checker-spot butterfly *Euphydryas editha*
Chimpanzee *Pan troglodytes*
Cod *Gadus morrhua*
Corn bunting *Miliaria calandra*
Corydoras catfish *Corydoras aeneus*
Cowpea weevil *Callosobruchus maculatus*
Damselfly *Calopteryx maculata*
Domestic fowl *Gallus gallus*
Drosophila bufurca
Drosophila melanogaster
Drosophila pseudo-obscura
Duck-billed platypus *Ornithorhynchus anatinus*
Dungfly, yellow, *Scatophaga stercoraria*
Dunnock *Prunella modularis*
Elephant seal *Mirounga angustirostris*
Fire ant *Solenopsis invicta*
Flatworm *Pseudoceros bifurcus*
Flycatcher, pied *Fiecdula hypoleuca*
Fruitfly, see *Drosophila*
Fulmar *Fulmarus glacialis*
Galapagos hawk *Buteo galapagoensis*
Garter snake *Thamnophis elegans*
Ghost spider crab *Inachis phalangium*
Giant squid *Architeuthis*
Giant water bug *Abedus herberti*
Goldfinch *Cardeulis cardeulis*
Gorilla *Gorilla gorilla*
Goshawk *Accipiter gentilis*
Great reed warbler *Acrocephalus arundinaceus*
Greenfinch *Careulis chloris*
Ground squirrel, thirteen-lined *Spermiophilus tridecemlineatus*
Grunion *Leuresthes tenuis*
Guinea pig *Cavia porcellus*
Gunnison's prairie dog *Cynomys gunnisoni*
Guppy *Poecilia reticulata*
Hamster *Mesocricetus auratus*
Harlequin beetle *Acrocinus longimanus*

Herring *Clupea harengus*
Hihi *Notiomystis cincta*
Honey bee *Apis melifera*
Honey buzzard *Buteo apivorus*
Honey possum *Tarsipes rostratus*
House fly *Musca domestica*
House sparrow *Passer domesticus*
Hyrax *Procavia* and *Heterohyrax*
Indigo bunting *Passerina cyanea*
Jacana *Jacana* species
Javan wart snake *Acrochordus javanicus*
Junco, dark-eyed *Junco hyemalis*
Kestrel *Falco tinnunculus*
Kiwi *Apteryx* species
Lake Eyre dragon *Ctenophorus maculosus*
Langur, Hanuman *Presbytis entellus*
Lion *Panthera leo*
Lungfish, Australian *Neoceratodus forsteri*
Magpie *Pica pica*
Mallard *Anas platyrhynchos*
Mantis shrimp *Pseudosquilla ciliata*
Marine iguana *Amblyrhynchus cristatus*
Noctule bat *Nyctalus noctula*
Opposum *Monodelphis domestica*
Opossum, Virginia *Didelphis virginiana*
Orang-utan *Pongo pygmaeus*
Ostrich *Struthio camelus*
Palolo worm *Eunice viridis*
Paper nautilus *Argonauta argo*
Peafowl *Pavo cristatus*
Pig *Sus scrofa*
Porcupine *Hystrix africaeaustralis*
Possum *Monodelphis domestica*
Prairie vole *Microtus ochrogaster*
Praying mantis, European *Mantis religiosa*
Pseudoscorpion *Cordylochernes scorpiodes*
Purple martin *Progne subis*
Redback spider *Latrodectus hasselti*
Red-billed gull *Larus novahollandiae*
Red-winged blackbird *Agelaius phoenicius*
Reed bunting *Emberiza schoeniclus*
Rove beetle *Aleochara curtula*
Sand lizard *Lacerta agilis*
Sand martin *Riparia riparia*
Scorpion fly *Panorpa vulgaris*
Screwworm fly *Cochlionmyia hominovorax*
Silk moth *Bombyx mori*
Skylark *Aluada arvensis*
Soay sheep *Ovis aries*
Spotted hyena *Crocuta crocuta*
Stick insect *Necroscia sparaxes*
Superb fairy wren *Malurus cyaneus*
Swallow *Hirundo rustica*
Turkey *Meleagris gallopavo*
Vasa parrot *Caracopsis vasa*
Viscacha *Lagostomus maximus*
Zebra finch *Taeniopygia guttata*

The sperm image at the head of each chapter is that of a Greenfinch *Carduelis chloris* drawn by Gustaf Retzius (Retzius, G. 1909. *Die Spermien der Vogel*).

Index

Numbers in italics indicate figures.